SCIENCE WARS

Politics, Gender, and Race

Anthony Walsh

Routledge
Taylor & Francis Group

LONDON AND NEW YORK

First published 2013 by Transaction Publishers

Published 2017 by Routledge
2 Park Square, Milton Park, Abingdon, Oxon OX14 4RN
711 Third Avenue, New York, NY 10017, USA

Routledge is an imprint of the Taylor & Francis Group, an informa business

Library of Congress Catalog Number: 2012031237

Library of Congress Cataloging-in-Publication Data

Walsh, Anthony, 1941-
Science wars : politics, gender, and race / Anthony Walsh.
    pages cm
  ISBN 978-1-4128-5163-3
    1. Science--Social aspects. I. Title.
Q175.5.W344 2013
303.48'3--dc23

                                                      2012031237

ISBN 13: 978-1-4128-5163-3 (hbk)

To my wife and soul mate, Grace; my children, Robert and Michael; my stepdaughters, Kasey and Heidi; my grandchildren, Robbie, Ryan, Mikey, Randy, Christopher, Ashlyn, Morgan, Stevie, and Vivien; and my great grandchildren, Kaelyn and Logan.

Contents

Preface

Few things titillate and entertain the intellectual palate more than the frequent iterations of the science wars, and perhaps no other issues goad academics in the social sciences to bellicosity more than gender and race when addressed in terms of biological differences. As Alun Anderson commented about these forbidden areas, "At a recent session I chaired at the World Economic Forum in Switzerland on Gender and the Brain, the real anger came from US scientists and intellectuals, venting their frustration that any discussion of biological differences relating to sex or race is a forbidden zone in universities in America" (http://www.edge.org/3rd_culture/leroi05/leroi05_index.html). While such discussions are not exactly forbidden, conducting research in these areas, or even commenting favorably on that research, can open a person to scorn and ostracism from colleagues, to student boycotts and harassment, and to having to endure those hissing epithets some folks are so fond of: sexist and racist.

When these issues have been addressed, they have tended to be in the form of philosophical debates, which are often less than polite, rather than as research questions. This book is both about that debate and about the research that has been done. Neuroscientists are increasingly using neuroimaging technology as it becomes ever cheaper to peer into the male and female brain, looking for structures and functions that underlie gendered behavior. This has been called neurosexism by scholars on the other side of the science barricade (Fine 2010). With the completion of the Genome project and the availability of ten-dollar cheek swabs for DNA collection, race—or as it is now often described in more politically palatable terms, population—is being conceptualized in molecular rather than in morphological terms.

This occurrence has led the enemies of such studies to opine that "biological racism" has replaced "phenotypic racism" (Amin 2010).

I have broadly divided the warring parties into biologists and social constructionists. Biologists tend to go about their business in their labs and offices, oblivious to the things social constructionists say about them—or even what a social constructionist is. Constructionists, however, tend to be acutely aware of what biologists are doing in the contested territory, although it is debatable whether they understand it. Of course, lobbing ad hominem racist and sexist grenades at the opposition is not the only tactic in the social constructionists' field manual. They may attack the very idea of science itself as a privileged way of coming to know the world, and they certainly stage frontal attacks on the concepts central to scientific thinking, such as determinism, reductionism, and essentialism.

The first chapter introduces the ideas motivating the contending parties in the science wars, beginning with Samuel Coleridge's notion that everyone is born either a Platonist or an Aristotelian. This provides a heuristic framework in which to explore the constructionist and biological positions on gender and race. What it means to be a Platonist or Aristotelian is briefly discussed and leads me to Thomas Sowell's constrained and unconstrained visions. These visions are essentially worldviews that map (imperfectly) to rightist and leftist views on many things. These visions are seen as the result of innate temperaments interacting with experiences to produce a framework for viewing the world. I then provide a short overview of the struggles of science with the church (Copernicus, etc.), Romanticism's reaction to the Enlightenment, and the reception of Darwin's theory of evolution. I end the chapter discussing the struggle for the souls of their disciplines between social scientists with conflicting visions.

Chapter 2 examines social constructionism, the issues it has with science, and its weak and strong versions. Everyone is a weak constructionist, for we have to admit that nature doesn't reveal herself to us packaged and labeled; humans must do it for her. However, strong constructionism wants to convince us that facts, not just the referents to such facts, are also socially constructed. It is strong constructionism that biology is battling when it denies the biological underpinnings of gender and race. I then examine what Ian Hacking calls the sticking points between science and constructionism on so many things—contingency, nominalism, and the stability of science.

Chapter 3 seeks to understand why there actually are people who are hostile to both the practice and the products of science. I look at Plato's rationalism and Aristotle's empiricism and conclude that they are often complementary. Francis Bacon's famous Four Idols that served as his arguments for empiricism are introduced, as is his frequent use of metaphors that radical constructionist feminists abhor as advocating the rape of women and which they use to discredit the scientific enterprise. Thomas Kuhn's work has been used (without his blessing) to support constructionist arguments about the instability, subjectivity, and relativism of science—arguments that I dismiss based on Kuhn's own work. Relativism is then given a broader treatment with the goal of understanding why it is so attractive to those with unconstrained visions.

Chapter 4 looks at the arguments over the conceptual tools that scientists tend to love and constructionists abhor—determinism, essentialism, and reductionism. I examine how scientists and their critics view these concepts, about which there are gradations of acceptance, and how each one of these concepts necessarily entails the other. Opponents tend to address only the most extreme versions of these concepts, as if they were sole representations, which leads them reject them completely. I respond by illustrating how useful the concepts have been to the advancement of science.

Chapter 5 asks one of the most fundamental questions we can ask about ourselves: What is human nature? All theories of human conduct contain an underlying vision of human nature, although these visions differ radically, and some claim that there is no such thing. I look at how natural selection has forged a sex-neutral human nature (the sum of our adaptations) in response to environmental challenges, and how sexual selection has forged separate male and female natures in response to sex-specific mating and parenting challenges. The chapter emphasizes the co-evolution of genes and culture and provides specific examples.

All chapters up to this point were devoted to the conceptual tools of science and the constructionists' objections to them. In Chapter 6 I take these sharpened and oiled tools into the substantive fray. I begin with gender, and how constructionists view it. Although gender is conceptualized separately from sex, biologists aver that sex and gender make a tightly knit bundle, while most constructionists want to deny any biological underpinning to gender. I examine the influence of anthropologists Margaret Mead and Melford Spiro, both of whom were strong cultural constructionists in their earlier careers but who later came to recognize the power of biology in sex/gender differences and in human nature in general.

Chapter 7 stresses that the neuroscience explanation of gender differences rests on a foundation of differential neurological organization, shaped by a complicated mélange of prenatal genetic and hormonal processes that reflect sex-specific evolutionary pressures. I discuss how the SRY gene begins the process of sexing the brain, and how this process can sometimes go awry. When it does, we see a variety of disorders of sex development that tell us a lot about normal gendered behavior. With respect to this issue, I also discuss transsexuals and chromosomal males with abnormal penis conditions. The sum of this evidence leads to the rejection of the notion of gender neutrality at birth, and of the implied constructionist notion that humans have somehow escaped their evolutionary history and have come to rely only on socialization to form gender identities.

Chapter 8 extends the previous one by going beyond the brain sexing process that occurs in utero. I respond to charges of neurosexism aimed at a neuroscience that is supposed to advance the position of a hardwired brain by showing that a central tenet of the discipline is brain plasticity, which is exactly the opposite of hardwiring. The emphasis is on how differences in brain structure and functioning, forged by different reproductive evolutionary pressures, are in evidence today in behavioral differences between the genders. Special attention is given to perhaps the largest behavioral difference of them all—criminal behavior. I aver that this difference is driven by different levels of fear and empathy (both higher in females), and that these differences reflect sex-differentiated parenting versus mating effort.

Chapter 9 moves from gender, which is a contentious issue, to race, which is positively incendiary. I look at how some have tried over the course of the 20th century to bury the concept of race, while others keep digging it up. There has certainly been an explosion of interest in race since the completion of the Genome Project. The idea of race has been re-conceptualized today in molecular terms, which pleases some and scares others. I look at the history of race and racism going back to ancient Greece and Islamic slavery, and inquire whether race is a socially dangerous idea. Racist stereotypes in the ancient world were every bit as nasty as, and perhaps even worse than, they are today, and I gainsay the idea that race and racism were inventions of the British and Americans to justify colonialism and slavery, respectively. I end with a look at racism in the modern United States.

Chapter 10 looks at major definitions of race—essentialist, taxonomic, population, and lineage—as they have changed from the time of the Enlightenment to the present. I look at how biologists traditionally defined subspecies (the 75% rule) and how physical anthropologists identify skeletal remains by race. I then look at how some biologists and psychologists explain race differences in behavioral and psychological traits with reference to the evolutionary life history theory, and how this type of research has been strongly attacked by those who feel that it can be used for racist ends.

Having examined race in terms of morphology, psychology, and behavior, I move in Chapter 11 to the level of molecular genetics, where some spectacular work is being done. The constructionist argument is that race does not exist as a biological entity, so I ask how one proves that something does not exist. Constructionists have put their faith in genetics to affirm their faith in the non-existence of race ever since a 1972 study by Richard Lewonton alleged that there is not enough genetic difference among human groups to support the idea of race. However, numerous studies have shown that geneticists can identify race with an accuracy approaching 100%, with relatively few gene variants.

The final chapter tries to bring the two sides together by briefly reiterating what they can each bring to the peace table. I urge constructionists to learn something about their enemy's strengths so they may be better prepared to engage them in the future. Ad hominem attacks and bald denials just won't cut it anymore. I explain the nature of arguments that arise from fear and moralistic fallacy in the hope that these forms of engagement will no longer be used, and I offer arguments that I believe constructionists can bring to the table in terms of race and gender now that the biologists' positions on these things seem almost unassailable. I also suggest that perhaps dropping the term race and substituting something more politically palatable might be in everyone's best interest.

Acknowledgments

I would first of all like to thank executive editor Mary Curtis for her faith in this project from the beginning. Thanks also for the commitment of her very able assistant Jeffrey Stetz. This tireless duo kept up a most-useful three-way dialogue between author, publisher, and excellent reviewers. The Transaction editorial staff spotted every errant comma, dangling participle, missing reference, and misspelled word in the manuscript—for which I am truly thankful—and Transaction's production editor, Bruce Meyers, made sure everything went quickly and smoothly thereafter.

I would also like to acknowledge the kind words and suggestions of those who reviewed this project. I have endeavored to respond to those suggestions and believe I have adequately done so. Any errors or misinformation that may lurk somewhere in these pages, however, are entirely my responsibility. Last but not least, I would like to acknowledge the love and support of my most wonderful and drop-dead gorgeous wife, Grace Jean, a.k.a. "Grace the Face." Grace's love and support have sustained me for so long that I cannot imagine life without her; she is my second sun, bringing me a world of warmth, light, love, and comfort.

Figure 1
Plato and Aristotle as depicted in Raphael's famous fresco
The School of Athens

Chapter 1

The Science Wars

Platonist or Aristotelian?

In Raphael's famous fresco *School of Athens* on the wall of the Stanza in the Vatican, the central figures are Plato, the incomparable master, and Aristotle, his star pupil. As physically close as these two ancient geniuses were in space and time, mentally they are archetypes of two radically different modes of philosophical thought. In the painting Plato is gesturing upward toward the heavens and Aristotle downward to the earth, symbolizing the central aspects of their respective philosophies and temperaments: Plato the top-down rationalist dreamer and Aristotle the bottom-up pragmatic empiricist. In the classical scholar A.T. D. Porteous's (1934:97) analysis of the philosophies of these two pillars of Western thought, he quotes 18th-century Romantic poet and philosopher, and definite Platonist, Samuel Taylor Coleridge:

> Every man is born an Aristotelian or a Platonist. I do not think it possible that anyone born an Aristotelian can become a Platonist; and I am sure that no born Platonist can ever change into an Aristotelian. They are two classes of man, beside which it is next to impossible to conceive a third. The one considers reason a quality or attribute; the other considers it a power.

Although such strict dichotomies are often suspect, Coleridge's division is a useful heuristic with which to examine two contrasting positions of the combatants in the science wars. Plato hungered for some solid and unchanging foundation on which to base his search for truth, which was a truth that had to transcend the world of the senses because the senses lead us astray. The reality of the phenomenal world was not denied, but he claimed that the objects we perceive are mere shadows mimicking the ultimate, which consisted of eternal and immutable "ideas" or "Forms" that transcended space and time and

1

existed in "a place beyond the heavens." While all the things that exist and are experienced in the phenomenal world participate in the Forms (how could we judge a person or object beautiful, or an act as unjust, if we did not have an archetype of the Forms of beauty and justice?), they are corrupt, ever-changing, and imperfect copies of their perfect and unchanging Forms. For Plato it was better to think philosophically about the world than to actually observe and measure it.

Although a committed follower of the theory of the Forms during his 20 years at the Athens Academy, Aristotle would later have none of this quasi-mysticism. Analytic thinkers like Plato rely on deductions from supposedly a priori truths and the force of persuasive language; synthetic thinkers like Aristotle aim "at nothing except precision and exactitude of thought and language" (Porteous 1934:101). Yet there is much to be gained from analytic and synthetic thinking when mixed in the right proportions. Porteous describes Plato's thought as "challenging and revolutionary," which points to an unrealized reality and brings to the table an "emotional quality" that he finds lacking in "Aristotle's dispassionate analysis." "Aristotle is the master of those who know, as Plato is of those who dream," writes Porteous (1934:105).

One of Plato's dreams outlined in the *Republic* was social and individual perfection, toward the attainment of which he was prepared to use deceitful tactics. Although Plato considered truth to be one of the main virtues, he believed that censorship and "noble lies" are required to build and maintain his ideal state. He imagined that philosopher kings and princes schooled to human perfection would rule his perfect society. Plato's ideas have dribbled down to some modern academic dreamers with their own visions of social perfection and how to achieve it, and of what kinds of knowledge and opinions relating to sensitive or dangerous issues are permissible both inside and outside academia.

Aristotle rejected Plato's utopian political theory as too far removed from the reality of human nature, and he knew enough about human nature that he would never trust a self-anointed intellectual elite convinced that they alone have the truth and the right, indeed the obligation, to coerce the rest of us in the "correct" direction.[1] He seemed to have anticipated the Roman poet Juvenal's trenchant question, "*Quis custodiet ipsos custodes?*" His concern for who will guard the guardians led Aristotle to the conclusion that the rulers as well as the ruled must be subservient to the law. This tension between dreamers of human and societal perfection and realists who fear the dangers of utopian thinking has been played out in every historical period. We should note, however, that in his last work (*Laws*) the mature Plato recognized that the state he had envisioned in the *Republic* was an impossibility given human nature, although he continued see it as the ideal state. The *Laws* reflected a new psychology of human nature posited by a dreamer who had been mugged by reality (Laks 1990).

Temperament and Visions

Similar to Coleridge, economist-philosopher Thomas Sowell (1987) posits that two contrasting visions of the world have shaped human thought about the same things throughout recorded history: the constrained and unconstrained visions. The constrained vision views human activities as constrained by a self-centered and largely unalterable human nature. The unconstrained vision views human nature as formed exclusively by culture and posits that it is perfectible. With Aristotle, constrained visionaries say, "This is how the world *is*," and with Plato, unconstrained visionaries say, "This is how the world *should be*." Sowell often uses the terms "gut level" and "instinct" to describe how these visions intrude into human thinking: "It is what we sense or feel *before* we have constructed any systematic reasoning that could be called a theory, much less deduced any specific consequences as hypotheses to be tested against evidence" (1987:14). The contrasting visions are brought to life in one sentence: "While believers in the unconstrained vision seek the special causes of war, poverty, and crime, believers in the constrained vision seek the special causes of peace, wealth, or a law-abiding society" (Sowell 1987:31). Note that this implies that unconstrained visionaries believe that war, poverty, and crime are aberrations to be explained, while constrained visionaries see these things as historically normal, although regrettable, and believe that what has to be understood is how to prevent them.[2]

Unconstrained visionaries tend to be the young and starry-eyed supporters of hope and change, or their older brothers and sisters who are discontented with the status quo and thirst to change it. Constrained visionaries are satisfied with the status quo, and with a wary eye on the failed utopias of the past, they counsel us to be careful what we hope for, because we just might get it. The optimism of the unconstrained leads them to focus on society as the source of individual problems such as crime and poverty because the more defects that can be placed on society, the more hope they see for the future. The pessimism of constrained theorists leads them to blame such problems on the defects of human nature and believe that society's task is to try to mitigate them.

Both constrained and unconstrained visionaries value fairness, but fairness is a concept saturated with contradictory notions; we all praise it but differ as to when its promise is fulfilled. Constrained visionaries view fairness as an equal-opportunity *process*—a non-discriminatory chance to play the game—that governments can attempt to guarantee by law. Unconstrained visionaries tend to view fairness as equality of *outcome*—all participants are winners—that no power on earth can guarantee.

With so many fundamental differences between Platonists and Aristotelians, it is clear that the prospect of a happy peace between the two positions faces formidable ideological barriers. These barriers are temperamental rather than intellectual because our temperaments have a lot to do with the information we deem worthy of our attention before we begin to ponder intellectually (Jost, Federico, & Napier 2009). Temperament has heritable components such as mood (happy/sad), sociability (introverted/extraverted), reactivity (calm/excitable), activity level (high/low), and affect (warm/cold) ranging from 0.40 to the 0.60s (Bouchard et al. 2003). Numerous studies have found the heritability of liberalism-conservatism—which map closely to the unconstrained-constrained visions—in the mid-0.50s (Bell, Schermer, & Vernon 2009), and neuroscientists are finding that political orientations are correlated with variant brain structures (Jost & Amadio 2011; Kanai et al. 2011).

Talk of genetic bases for such things as political attitudes does not sit well with unconstrained theorists, who believe that we get our politics where we get our porridge: at the kitchen table. Of course, geneticists do not expect to find genes "for" an Aristotelian or Platonic worldview or a Sowellian vision by rummaging around among our chromosomes, nor do neuroscientists expect to see red and blue clusters of neurons in our brains. Our worldviews (visions) are synthesized genetically via our temperaments, which serve as physiological substrates guiding and shaping our environmental experiences in ways that increase the likelihood of developing traits and attitudes that color our world in hues most congenial to our natures (Olson, Vernon, & Harris 2001; Smith et al. 2011).

If genes account for between 40% to 60% of the variance in temperamental sub-traits, the environment accounts for the remaining variance. Thus, while our visions are resistant to change, they are not impossible to change. The notion of variance alerts us that the Platonic/Aristotelian, constrained/unconstrained dimensions are continua along which people shift back and forth according to the issue at hand, and are certainly not rigid dichotomies. Only a few souls are glued tightly to the tails of the distribution; these are the few dogmatic fundamentalists who infect any system of thought. The point is that temperament trumps reason in so many ways that matter. If it did not, we would not see such eminently *reasonable* thinkers as Plato and Aristotle, and their modern counterparts, differing so widely on important issues.

Early Science Wars: A Very Short History

This division of human temperaments along ideological fault lines helps us to understand why there have always been struggles between the visions of *is* and *ought,* and why there have always been arguments about the nature of knowledge, how it is to be acquired, and what aspects of it are permitted to see the light of day. We are all familiar with the church's self-declared role as the ultimate authority on truth, both temporal and spiritual, in days of yore. Secular knowledge was acceptable as long as it was supportive, or at least not contradictive, of church doctrine. Among the knowledge enjoying the church's imprimatur was the geocentric theory of the solar system, which claimed that the earth is the center of everything, and that the sun revolved around it. The church supported the geocentric model because it was consistent with biblical accounts, and because it placed the earth and human beings at the center of everything. Psalm 104:5 states that "He set the earth on its foundation; it can never be moved." And in Ecclesiastes 1.5, the writer says, "The sun rises and the sun sets, and hurries back to where is rises." The geocentric theory also comports with our immediate sense experiences. We don't feel the earth moving as it spins on its axis at just over 1,000 miles per hour while hurtling through space at about 67,000 per hour, and we do see the sun rise in the east, move across the sky, and then set in the west; how could any sensible person not be a geocentrist?

Then along came the Polish astronomer, priest, and polymath Nicholas Copernicus with his heliocentric model of the solar system. This model not only defied common sense, but it also upset humankind's privileged position as central to the nature of things. He wisely did not trumpet his theory, since being politically incorrect in those days not only destroyed reputations, it could result in imprisonment, torture, and execution. Thus, his book announcing the model was not published until after his death in 1543. Copernicus also feared ridicule from fellow scientists, the vast majority of whom subscribed to the geocentric model on scientific grounds. Aristotle, considered at that time by most scientists to be the ultimate authority on science, had supposedly refuted heliocentricity, a refutation held by almost all Copernicus's contemporaries as self-evident (Eichner 1982). So it is not always, or not even mostly, opponents of science, but scientists themselves who may oppose scientific progress if it happens to threaten long-held treasured positions in which they may have a strong emotional investment.

Copernicus's prudence meant that the brunt of the battle for the heavens was to be borne by others. Friar/philosopher Giordano Bruno was burned at the stake in 1600 for Copernican heresy, along with an assortment of other heresies both spiritual and temporal. But the man who really brought the issue to a head was Galileo Galilei, a man both Albert Einstein and Steven Hawking credit with being the father of modern science (Radhakrishna 2009). The details of his battles with the Church need not concern us, but he did publish his book (*Dialogue Concerning the Two Chief World Systems)* during his lifetime and was duly convicted of heresy. He was sentenced to prison, a sentence later commuted to house arrest. He died in his home in 1642, nine years after the Inquisition had condemned him.

These clashes between science and religion over this and other matters harmed both and benefited neither. It harmed science by delaying its progress, and it harmed religion by stamping it as stubbornly dogmatic and irrational. The church should have listened to St. Augustine, its brightest intellect, when he wrote that in "matters that are so obscure and far beyond our vision we should not rush in headlong and so firmly take our stand on one side that, if further progress in the search for truth justly undermines this position, we too fall with it" (in Collins 2005:83). Although Augustine was referring to scriptural squabbles, his statement applies far more to scientific quarrels, because only science can promise "further progress in the search for truth"; only science has the tools to do so.

The science battles were considerably more civil and were conducted more eloquently in the 19th century, when the tension was between the conservative elders of the Enlightenment and the rebellious children of Romanticism. Romanticism began around 1800 as a counter to what many humanists considered a cold and mechanistic science (Aristotle's "dispassionate analysis") that had disengaged humanity from nature. Rather like contemporary postmodernists, the new generation of intellectuals valued creativity and spontaneity, and saw science with all its rules and methods as a numbing rationality lacking in Plato's "emotional quality." The quarrel was not so much with science per se, but with disembodied, emotionless, reductionist science that ignored the philosophical, spiritual, and ethical implications of its works. But perhaps nature only becomes accessible when she is alienated from human passions and desires. Romanticism's particular distaste for reductionism is clear in William Wordsworth's famous poem "The Tables Turned," supposedly written in response to Sir Isaac Newton's explanation of the rainbow in terms of the soulless physics of refracted light:

> Sweet is the lore which Nature brings;
> Our meddling intellect
> Misshapes the beauteous forms of things
> We murder to dissect.

In the final stanza of the poem Wordsworth cries "Enough of science," which is also the sentiment expressed in horrific metaphor in Mary Shelley's book *Frankenstein*. It too was a revolt against reductionism and the gross manipulation of nature. Dr. Frankenstein's monster was an assemblage of body parts gathered from midnight raids in dissecting rooms and cobbled together without any consideration of the consequences to either the monster or to the community upon which it was to be foisted. Shelly was saying that science could produce monsters with an apparent lack of concern for the whole and with a cold interest in disengaged parts, urging that it must do more to appreciate nature and less to subjugate it (Bentley 2005).

Anyone applying biological science to the social science rainbow is likewise sure to be met with similar but less eloquent objections, for reductionism is one of social science's favourite boo words. But it is not well known that Wordsworth eventually came to realize that rainbows are no less beautiful when understood as refracted light, and he became an ardent admirer of Newton. In one of his *Isle of Man* sonnets, he expressed the opinion that we should not try to "hide Truths whose thick veil Science has drawn aside" (in Jeffrey 1967:21).

Having been dethroned from its central position in the solar system by Copernicus, humanity was to receive another massive body blow from Charles Darwin, whose theory of evolution stamped humans as just another animal forged by nature rather than divine creativity.[3] Darwin was very much a romantic who saw beauty in lowly earth-worms as well as stately palm trees, but he was also a meticulous scientist. His obsessive collecting while on expedition aboard H.M.S *Beagle* led him to the elegant theory of natural selection. Although Darwin formulated his theory from a multitude of facts, it was considered highly speculative and lacking in predictive value. Positivism was much in vogue at that time. Positivism is a doctrine maintaining that science should concern itself only with phenomena that can be directly observed, and it is hostile to theoretical speculation (natural selection was speculative in the sense that its deeper genetic mechanisms were not known at the time). Because of this, many scientists rejected the theory, but just as many embraced it. There were some snide remarks, such as inquiries as to which side of his family Darwin claimed descended from monkeys, and the theory was condemned from many pulpits. Yet he was never threatened by inquisitions or physically attacked by secular opponents, as some of today's politically incorrect scholars have been physically attacked and otherwise harassed by modern zealots. The esteem in which Darwin was held can be gauged by his final resting place at Westminster Abbey (alongside Newton), a lofty honor reserved only for Britain's giants of philosophy, literature, statecraft, warfare, and science.

It is difficult for us to understand today how humiliating and offensive Darwinism was for so many people, and how thoroughly destructive many thought it might be to the social order. Darwin himself was deeply concerned with this possibility. The wife of Bishop Samuel Wilberforce is said to have remarked about the theory, "Let us hope it is not true, but if it is, let us pray it will not become generally known" (Kliman & Johnson 2005:926). Note that while horrified by the theory and its presumed consequences if "generally known," Mrs. Wilberforce did not categorically reject it or impugn the characters and motives of those who supported it. We should all likewise acknowledge that many things we may not like and find offensive, and hope not to be generally known, may indeed be true. We should also with Wordsworth not seek to hide those truths "whose thick veil Science has drawn aside."

The Current Scene

When I was an undergraduate in the early 1970s, the intellectual landscape was fairly clear and unambiguous—or at least I thought it was. The two main divisions were the humanities and the sciences, inhabited by smart and soft-hearted romantic Platonists in the first case and by the *really* smart and hard-headed Aristotelians in the second. Each division knew its place in the order of things; the sciences told us to look inside our heads and explore how the world operates—how the world is—and the humanities implored us to look into our hearts and image how the world ought to or could be. The stereotypical division was between the practical and the beautiful, the in-your-face objectivists and the in-your-heart subjectivists. The caricature of science was that it was a dull laboratory plod wherein resided anally meticulous geeks with bifocals and pocket calculators who would call Hamlet's "To be or not to be" a tautology, and whom the poet e. e. cummings described as a "oneeyed son of a bitch [who] invents an instrument to measure Spring." The caricature of the humanities was that it was peopled by the clever but not brilliant flower children in ponytails and sandals who could turn a nice phrase or paint a pretty picture, but who were too math phobic and undisciplined to master science. Nevertheless, they brought to the table beauty and a different way of looking at things.

There were no "wars" between C. P. Snow's two cultures (Snow comfortably straddled both) in the middle of the 20th century, usually just indifference. The scientific and humanities communities were too far apart to declare war and shared no common ground on which they could engage one another.[4] There are disciplines such as anthropology, psychology, and sociology, however, that are not sure in which camp they should fly their colors. Are they sciences, or are they social philosophy dressed up as sciences? There are sometimes acrimonious debates—we might call them "civil wars"—within these disciplines about this question, with some opting for science and others for such things as postmodernism. Anthropology departments across the country are fracturing, and even formally separating, with physical anthropologists embracing science and many cultural anthropologists opting to take a dive off the postmodernist cliff. The stinging criticism of social science by anthropologist John Tooby and psychologist Leda Cosmides (1992:23; emphasis added) makes no bones about where their disciplines should be flying their flags:

> After more than a century, the social sciences are still adrift, with an enormous mass of half-digested observations, a not inconsiderable body of empirical generalizations, and a contradictory stew of ungrounded, middle-level theories expressed in a babble

of incommensurate technical lexicons. This is accompanied by a growing malaise, so that *the single largest trend is toward rejecting the scientific enterprise as it applies to humans.*

Then there is the book *What's Wrong with Sociology?* (Cole 2001), containing contributions from some of the brightest stars in sociology. For almost 400 pages these authors collectively bemoan the state of their discipline, complaining about its saturation with left-wing ideology, the mundane nature of its studies, its methodological faddism, and its lack of scientific rigor. But this is the contested point: what are the social sciences, and what is their mission? Some say that the mission is not at all to ape the natural sciences and objectively observe and measure phenomena in Aristotelian fashion, but rather to follow Plato and think about phenomena and then use the insights obtained to seize the moral high ground. The goal of those in this camp is not necessarily to understand, but to serve as critics of the status quo and act as the collective conscience of society. These people have added a third culture to Snow's two and seek to impose on both cultures the dogma that there are no objective, natural, and universal truths—of this they are absolutely sure.

Anthropologist Charles Leslie modeled the behavior expected of the anointed ones in the social science when he resigned his editorial position at the journal *Social Science and Medicine* in a self-righteous huff because it published an article on AIDS that he considered racist (it documented the huge overrepresentation of Africans or people of African descent among people with AIDS). Leslie (1990:896) tells us with refreshing honesty that he does not want objective science to infect social science and thus divert it from its true mission: "Nonsocial scientists generally recognize the fact that the social sciences are mostly ideological. . . . Our claim to be scientific is one of the main academic scandals. . . . By and large, we believe in, and our social science was meant to promote, pluralism and democracy." I assume that most of us like Leslie's pluralism and democracy, but he wants to support only his vision of what these things mean, even if it means condemning to oblivion work on something as important as AIDS research because it violates his worldview. This is hardly pluralistic or democratic, but then, Leslie evidently believes in the righteousness of Plato's concept of noble lies, even when used to conceal scientific data. He sees social scientists as Platonic guardians whose role is social criticism and officiating at the burial of discomforting data they don't want to be generally known, not the seeking of a firmer understanding of human nature.

Most practitioners of the natural sciences ignore the so-called wars, confident that their laboratories are quite safe from any firecrackers the postmodern crowd may lob at them. Most practitioners of the humanities are likewise unworried and can experiment to their heart's content with all kinds of radical ways of thinking and seeing, because it's all subjective, anyway. For social scientists, however, the wars are battles for the souls of their disciplines. With social scientists straddling both camps, it was certain there would be some skirmishes between the more bellicose types. Most of the initial pinpricks came from scholars associated with the sociology of science, particularly from the French intellectual left personified by figures such as Baudrillard, Foucault, Latour, and Lacan. The radical positions vis-à-vis science, objectivity, and the status quo of these thinkers appealed greatly to American leftists in the humanities and social scientists, and their *haute culture* French pedigree didn't hurt, either.

The opening salvo of the present iteration of the science wars proper was arguably fired by biologist Paul Gross and mathematician Norman Levitt, two scientists who took the anti-science people seriously and who were appalled by the silliness of many of the positions they supported. Gross and Levitt's scorching book *Higher Superstition: The Academic Left and Its Quarrels with Science* (1994) excoriated strands of the humanities and social sciences that attack the deeply unfashionable view that the world actually exists independently of perspectives, prejudices, and power relations that exist at a particular time and place.

Higher Superstition was a straightforward frontal attack that drew counterattacks on the academic battlefield but did not touch the civilian population. On the other hand, Alan Sokal's parody of postmodernist thinking, published in the cultural studies journal *Social Text* as a serious contribution, got the attention of intellectual types both inside and outside academia. Sokal's article, impeccably adorned with impenetrable postmodernist prose and entitled "Transgressing the boundaries: Toward a transformative hermeneutics of quantum gravity" (1996), was a Trojan horse snuck into the postmodern camp as a gift from a "real scientist" (a physicist) apparently endorsing its semi-solipsistic view of the world. Sokal revealed his hoax in the now defunct American literary magazine *Lingua Franca* (1996), later describing his article as "a mélange of truths, half-truths, quarter truths, falsehoods, non sequiturs, and syntactically correct sentences *that have no meaning at all*" (Sokal & Bickmont 1998:208–209; emphasis added). The article purported to demonstrate that quantum gravity was at bottom a social construct with political implications, and that this immensely complicated field that attempts to unify quantum mechanics and general relativity in a "theory of everything" is "clearly . . . an archetypal postmodernist science" (Sokal & Bickmont 1996:234).

Sokal's hoax stung, and his targets, unable to distinguish between the meaningful and the meaningless, whined like losers in the locker room about their opponent's perfidious tactics. But Sokal was no neo-conservative lambasting the loony left. As a self-described "old time leftist," his stated aim was to rescue the left from what he considered the scientific irrationality into which it has descended. Peter Singer made a similar (but serious) effort to rescue his beloved left from what he considered its irrational utopianism and its rabid anti-naturalism in his book *A Darwinian Left: Politics, Evolution, and Cooperation* (2000). In this book, Singer wanted to present "a sharply deflated vision of the left, its utopian ideals replaced by a coolly realistic view of what can be achieved" (2000:62). Singer maintains that among other things the left must accept the reality of human nature and seek to understand it as natural science (particularly evolutionary biology) does, and it must stop assuming that all inequalities are due to prejudice, discrimination, and oppression, which seems to be the assumption of the bulk of social scientists.

The left, of course, has traditionally denied that there is such a thing as human nature, because of the biological undertones accompanying that concept, but the left really has nothing to fear. In fact, Peter Grosvenor (2002:446) opines that "the intellectual left is likely to be the prime beneficiary [of a Darwinian worldview] if the social sciences and the humanities can be rescued from residual Marxism and obscurantist postmodernism." Needless to say, Marxists and postmodernists do not feel the need to be "rescued." But what exactly is the nature of the enemy that Sokal and Singer identify, and why is it of importance to social scientists with Aristotelian rather than Platonic temperaments? With sincere apologies to social constructionists to whom this does not apply, in general terms the enemies of science may be lumped under the broad umbrella of social constructionism, a movement addressed in the next chapter.

Conclusion

The so-called science wars have raged with various levels of intensity ever since humans began thinking about their world and tried to understand it. The ultimate source of these wars probably lies in different human temperaments that move our thoughts on so many things in divergent directions. We call these positions by various names that only partially capture the complexity involved, such as Platonist/Aristotelian, unconstrained/constrained visionaries, liberal/conservative, soft-hearted/hard-headed, romantic/realist, and so on. These positions certainly overlap; many sociopolitical liberals are hard-headed scientists, and many sociopolitical conservatives are ardent Platonists. The intellectual fracture, however ragged it may be, generates conflict, and that's a good thing, because conflict stirs the water and keeps it from stagnating. Criticism, attack, and counterattack, keeps the mind hopping, healthy, and alert—and let's face it, it is jolly good fun.

Ian Hacking (1999:62) puts something of a curse on the houses of both sides of the conflict: "The science wars, as I see them, combine irreverent metaphysics and the rage against reason on one side, and scientific metaphysics and an Enlightenment faith in reason, on the other." Hacking sees metaphysical issues as central to both positions. I doubt whether social constructionists will accept Hacking's characterization of them as raging against reason, or that scientists will accept Hacking's intimation that their faith in science is too reverential. Because Hacking is a central philosophical figure in the issues at hand, and he considers himself more dispassionate umpire than participant, it might be wise to accept his assessment that the wars are metaphysical in nature, resting on temperament-driven visions for their truths, rather than on substance.

Endnotes

1. Like all great thinkers, Plato could contradict himself. His utopian vision showed an incredible misunderstanding of human nature (as do all such visions). On the other hand, his Ring of Gyges allegory (suggesting that conscience is a mechanism of morality engaged only when others are looking) is perhaps the best short précis on human nature available, and with which people (at least my students) connect instantly. His *Laws*, of course, indicates that he eventually gave up on utopian visions of human perfectibility.
2. The flavor of Sowell's argument is better perceived with a few examples of the thinkers he placed in each vision. In my estimation, thinkers in the constrained tradition include Adam Smith, Thomas Hobbes, Edmund

Burke, Alexis de Tocqueville, Oliver Wendell Holmes, F. A. Hayek, and Milton Friedman. More contemporary examples are Ronald Reagan and Margaret Thatcher. In the unconstrained corner are Jean-Jacques Rousseau, Thomas Paine, William Godwin, Condorcet, Harold Laski, and John Kenneth Galbraith, with Jimmy Carter and Barack Obama being more modern examples.

3. Plato's *Republic* contains a dialog pertaining to artificial selection, and it even hints that nature might act in the same selective manner on humans (Plato 1956: book V).

4. Snow (1964:74) suggested that molecular biology (genetics) would eventually provide a way to bridge the chasm between the natural and social/behavioral sciences because "it is likely to affect the way men [and women, of course] think of themselves more profoundly than any other scientific advance since Darwin." I believe that he was right, and I have published works attempting to show how it is possible in criminology (Walsh 2009).

Chapter 2

Social Constructionism

What Is Social Constructionism?

Social constructionism and its sibling social constructivism are often used interchangeably, although they refer to different but greatly overlapping processes. Constructivism is usually taken to be the process by which individuals generate subjective meaning from the knowledge they receive, and it is a concept associated primarily with Piagetian ideas about the psychology of developmental learning. Social constructionism, on the other hand, is a sociological theory of knowledge (loosely defined, for it is more a mode of critique than a theory from which we can derive hypotheses) maintaining that concepts, practices, beliefs, and sometimes facts are artifacts of a particular time and place. These artifacts (constructions) are said to be contingent on human representations for their existence rather than on some inherent property those things possess. Because human beings construct their individual meanings in a social context, they all share in the constructs of their cultures to various extents, thus the constructivist/constructionist distinction is not particularly helpful for the present purpose. Therefore, I will use constructionism throughout except when quoting someone who uses constructivism as a synonym.

In his book *The Social Construction of What?*, Ian Hacking, a big gun in the philosophy of science, catalogs at least fifty tangible and intangible things, ranging from mental illness to quarks, that someone or another has claimed to be social constructs. We could all add multiple items to Hacking's laundry list of things said to be created from "social stuff." The ubiquity of social constructionist thought in the humanities and social science has even led to claims that it is a major contender for a metatheory of the social sciences (Gergen 1988). Although social constructionism is more an ontological-epistemological critique than

15

an explanatory theory, such a prospect would make any good Aristotelian social scientist shudder.

Constructionism may be social science's equivalent of Daniel Dennett's (1995) "universal acid" metaphor for Darwin's theory of evolution by natural selection. Dennett's point is that the logic of Darwinism is so overwhelming that it eats its way through every theory and concept relating to the living world that came before it, leaving behind a very different intellectual worldview in its wake. Unlike social constructionism, Darwinism is so obviously a theory with such solid foundations that "Nothing in biology makes sense except in the light of evolution" (Dobzhansky 1973:125). For their part, many social scientists on the science side of the wars are enthusiastic about Darwinism (in the form of evolutionary psychology) as a metatheory "to tie together the forest of hypotheses about human behavior now out there" (de Waal 2002:187). The prospect of "biologizing" their disciplines is equally likely to make a good Platonic social constructionist shudder.

Social constructionism may not be as irresistible as Darwinian acid, but perhaps we can view it as something softer, like a universal sponge, soaking up every traditional concept in the social and scientific worlds from A to Z and squeezing them back out in mutated form. Sponges are useful absorbers of spills (read, careless assumptions about what is really real, natural, and inevitable) and cleaners of impervious surfaces (read: rigid, obdurate ideas about the nature of reality). But sponges also provide mediums for the growth of harmful spores when slipshod users allow them to remain wet between uses. For all the acknowledged usefulness of the constructionist sponge, many constructionist ideas start out so wet that they cannot dry, thus giving opponents (such as Alan Sokal) every opportunity to poke fun at the absurdities that all too often spawn in their pores.

So what is social constructionism? Like every term ending with *ism,* there are many varieties that often conflict with one another, making it difficult to compose a definition to everyone's liking. Rather than engaging dueling dictionaries, I offer my own definition, cobbled together from a variety of other definitions: Social constructionism is a sociological model that emphasizes the socially created nature of truth and knowledge (and sometimes facts), and serves as an ontological-epistemological critique opposed to realism, reductionism, determinism, and essentialism. It is thus also a highly ideological model that tends to be popular with left-wing unconstrained visionaries. It is a blood-warming war cry, a battle slogan for those who want to free the oppressed and change the status quo. Constructionism tells these people that because humans have constructed things the way they are, humans can deconstruct them.

I rely largely on Hacking to explore social constructionism, because, as we recall from Chapter 1, he is not a combatant in the science wars. He sees himself as a neutral foreign correspondent simply reporting on the action and occasionally trying to act as a peacemaker. According to Hacking (1999:6), social constructionists hold that:

(1) X need not have existed, or need not be at all as it is. X, or X as it is at present, is not determined by the nature of things; it is not inevitable.

Very often they go further, and urge that:

(2) X is quite bad as it is.
(3) We would be much better off if X were done away with, or at least radically transformed.

Hacking breaks this general schema into six grades or degrees of commitment to constructionism, ranging from the extremely weak to the extremely strong. The *historical* grade is the weakest form; it refrains from judging X (any idea, thing, event, or institution) and merely claims that X is the result of certain historical events and could have been otherwise. The *ironic* form points out that X could have been otherwise, and those who believe X is inevitable are naïve and wrong, but we are stuck with it. The *unmasking* constructionist exposes the ideology and power relations that underlie X, but does so as an intellectual exercise only. The unmasker does not necessarily seek to refute X, only to expose it. The *reformist* approach takes a negative view of X as it is and seeks to modify aspects of it by pointing out the contingency of the aspects of X that he or she dislikes. The *rebellious* form urges that we would be better off without X altogether, and *revolutionary* constructionists become activists in the cause of doing away with X.

In this book I substitute gender and race for Hacking's generic Xs. Many subjects denoted by these labels have been treated badly over the centuries, and thus it is reasonable that many should be uneasy about certain so-called truths that have been attached to them and want to do something about it. In the first instance, the goal is to decouple gender from biological sex, and in the second it is to do away with the concept of race altogether. Each of these goals describes a strong opposition to biological thinking, although snippets of biological knowledge believed to comport with the constructionist position will be accepted gladly. I view social constructionism as being concerned primarily with deconstructing the truths our cultures have held to be factual and immutable about various features pertaining to these concepts, and biology as going about its business only dimly aware of constructionist claims. There are scholars who straddle both camps and who would like to see the two sides come together, but diehard constructionists will filibuster to their graves to prevent such a happening. As physicist Max Planck is supposed to have said about the intransience of older physicists regarding acceptance of quantum mechanics, "Science progresses one funeral at a time."

Weak versus Strong Social Constructionism

I follow convention and dichotomize Hacking's continuum into weak (his first three positions) and strong (the remaining three) versions. Because we cannot deny that the reality we perceive and act on is constructed from common experience and is communally validated, we are all weak social constructionists. Steven Pinker (2002:202) provides many examples of things that are obvious social constructions and that "exist only because people tacitly agree to act as if they exist. Examples include money, tenure, citizenship, decorations for bravery, and the presidency of the United States." These things and practices are no less real for being socially constructed, but the constructionist point is that they are not products of nature and are therefore contingent rather than inevitable. They may be ontologically objective in that they produce real consequences, but they are also ontologically subjective because they require human input in order to exist.

It could be argued that while we can get by without tenure and bravery decorations, perhaps it is inevitable that once organized into groups, and competing and trading with other groups, humans necessarily had to develop formal systems of leadership and efficient barter lest they descend into Hobbesian chaos. Inevitable or not, they are still human products rather than products of nature. Taking this reasoning a little further, at one level all things, including the gifts of nature, are socially constructed. Nature does not reveal herself to us already sorted and labeled, so humans must do it for her. Social construction in this weak sense means that humans have perceived a phenomenon, named it, and categorized it according to some taxonomical rule (also socially constructed) that takes note of similarities and differences among the things being classified. But because something is *necessarily* socially constructed in this vacuous sense, it does not mean that the process of categorization is arbitrary and without real empirical referents and rational meaning, as claimed by many who try to deconstruct concepts they dislike. We must not confuse the socially constructed referents to these things with the things themselves. There are problems with all social constructs, particularly the concepts addressed in this book, but then very few concepts in any domain of knowledge, with the possible exception of mathematics, are defined and understood in such a way as to make every application of their descriptors universally acceptable.

I have had a liking for the weaker strands of social constructionism ever since I read Berger and Luckmann's *The Social Construction of Reality* (1966) as an undergraduate. I still have my much-annotated and dog-eared copy. Although almost everything I read in the book became suddenly obvious as soon as I read it, it still made an impression on me because it poked me in the eye with its obviousness. So many things Berger and Luckmann addressed were already so obviously in the constructionist fold (such as Pinker's examples given above) that it now seems superfluous to have mentioned them at all, but at the time their unmasking was revelatory and enlightening. Like a fish yanked suddenly from the pond who appreciates water for the first time, I came to appreciate the tenuousness of social reality. Social constructionism can thus be liberating if it is not stretched to a breaking point.

Although the idea of social constructionism had been around for centuries, and perhaps most succinctly stated in the Thomas Theorem ("If men define situations as real they are real in their consequences"), the phrase itself and the diverse ideas it envelops arguably took on their present form beginning with this book. Berger and Luckmann were not radical strong constructionists in the sense that they believed everything is arbitrarily socially created, for they often stressed the biological substrates of many social constructions. Their main point was that people interact with the subconscious understanding that their perceptions of reality are shared with others, and when they act on this understanding, their faith in their reality is further reinforced.

Berger and Luckmann's only agenda was to stop us taking our realities for granted and reifying them, and to problematize what was previously unproblematic. Nothing in their book seemed to be claims about the existence of things, but rather about how those things are represented. After all, there must be something "out there" with substance that influences our referents and that is not created by those referents. As obvious as this is to most of us, there are serious doubters among a coterie of academics so lost in the abstractions that they rarely engage the real. Bruce Charlton, editor-in-chief of the journal *Medical Hypothesis*, takes a medical and evolutionary scalpel to such people, whom he calls "clever sillies." A clever silly state is a:

> somewhat tragic state; because it entails being cognitively trapped by compulsive abstraction; unable to engage directly and spontaneously with what most humans have traditionally regarded as psycho-social reality; disbarred from the common experience of humankind and instead cut-adrift on the surface of a glittering but shallow ocean of novelties: none of which can ever truly convince or satisfy. It is to be alienated from the world; and to find no stable meaning of life that is solidly underpinned by emotional conviction. (2009:869)

Charlton claims that clever sillies overthink everything to the point of utter confusion in matters where common sense would better suffice. They then go on to publish these garbled thoughts without suffering the negative consequences that professionals in applied fields outside of academia must endure for their mistakes. Engineers must build bridges that stand up, surgeons must save more lives than they lose, and business CEOs must keep their companies solvent and their stockholders happy. Such people have strong reality checks against intellectual recklessness, but in areas of academia where Charlton's sillies are most likely to reside, no reality checks are ever issued that they can cash. That wonderful thing called tenure frees us academics almost completely from normal restraints, which results in a tendency to take ideas beyond their logical extremes. This is often desirable for scientists struggling with the deep arcana of theoretical physics, but is often disastrous for those who deal with the everyday world of common sense.

Fact Constuctionism

Fact constructionism is an excellent example of what Charlton rails against. Ron Mellon asserts that few of us are surprised or disagree with weak constructionist claims: "While it is quite surprising to think that putatively natural phenomena like sex or race or quarks are the result of our culture or decisions, it is not nearly surprising to think that our theories and beliefs about these and other phenomena vary sharply from culture to culture" (2007:97). Constructionists of the strong variety do extend their constructionist reasoning to include things that many, most, or all scientists consider natural, such as Mellon's sex, race, and quarks. They maintain that our vision of reality in all domains, including the scientific domain, is simply discourse rooted in social consensus. If the social consensus is that X is real, it is; if the consensus is that X is not real, it is not. This is true of so many things for which there never was any empirical evidence (witches, gods, fairies, phrenology), even for things for which there was some, albeit flawed evidence (geocentrism, phlogiston, luminiferous ether). Strong constructionists, however, have upheld their position in matters for which there is unequivocal evidence coming from different disciplines using many different tools and methodologies.

In his book *Fear of Knowledge: Against Relativism and Constructivism,* Paul Boghossian, a physicist turned philosopher, distinguishes three forms of constructivism: truth, knowledge, and fact. He claims that the most influential of these forms is fact constructivism, which, like Mellon, he finds "somewhat surprising given that it is the most radical and the most counterintuitive." He goes on to say that "Indeed, properly understood, fact-constructivism is such a bizarre view that it is hard to believe that anyone actual endorses it. And yet, it seems that many do" (2006:25). I assume that most of us agree that there are objective facts that are products of nature (atoms, giraffes, mountains, viruses) that exist outside of human perceptions of them; that is, they exist whether or not humans are aware of them. Fact constructionists, however, aver that *any* fact obtains only because humans have constructed them based on the needs and interests of those who constructed them. For them, science laboratories are construction sites for facts, and no alleged fact about the world is autonomous of human action. This view promotes humans to godlike creatures who literally create reality and affirms the opinion of 5th-century Greek sophist Protagoras that "Man is the measure of all things, of those that are, that they are, and of those that are not, that they are not" (in Burnyeat 1976:44).

Sokal and Bricmont (1998:96–97) provide us with a quite bizarre example of fact constructionism from French sociologist Bruno Latour. When scientists working on the mummy of Ramses II concluded that Ramses died (in 1213 BCE) of tuberculosis, Latour commented "How could he pass away due to a bacillus discovered by Robert Koch in 1882? Before Koch, the bacillus had no real existence." (Latour slips here. His use of the term discovered implies something pre-existing). Sokal and Bricmont went on to say that Latour noted that it would be just as anachronistic to claim that Ramses had died from machine gun fire as it would be to claim that he died of tuberculosis! Latour slides between something that is obviously a thing of human manufacture to something that existed in the natural world long before Ramses II, and then claims their equivalence. The bacillus was certainly "called into being" as an answer to questions Koch and others had raised about the natural world, but to say that it didn't exist prior to Koch is just as absurd as saying that the earth didn't revolve around the sun before Copernicus.[1]

Not all examples of fact constructionism descend to this level of absurdity. Many of the constructionist arguments about humankind revolve around how we (humans in a variety of societies) have concocted facts according the interests and power relations obtaining at the time of construction. The putative social dimensions of fact constructionism are different from those of truth and knowledge constructionism. A strong constructionist will aver that there is no superior truth or knowledge, only different truths or knowledge. The validity of this claim jumps around according to what we are talking about. Scientists maintain that there is knowledge and there are truths arrived at through procedures of rational assessment called the scientific method, but they will agree that there are domains that their methods do not touch, such as history, morality, ethics, and aesthetics, where truth does depend to varying extents on non-evidentiary criteria such as politics, social consensus, individual temperament, and so forth.

However, strong constructions pull into their sphere of interest areas in which science does intrude, particularly where it intrudes in areas of cultural and behavioral relevance. Because many (perhaps most) practitioners in humanities and social sciences claim such areas as exclusively their domain, they raise strong objections when geneticists, neuroscientists, and evolutionary biologists bring their big guns into the fray. Other social scientists welcome these natural scientists as robust allies who can move the cause of human understanding forward. There are thus many bones of contention lying in the academic feeding pit. Hacking identifies three of them that he calls sticking points.

Sticking Points: Contingency, Nominalism, and Scientific Stability

Hacking sees these sticking points of disagreement to be so fundamental that he fears that the warring sides will be stalemated for a long time (after all, it has been the case at least since Plato and Aristotle).

He yearns for both sides to find some common ground from which they can settle matters if only they can only surmount the metaphysical barriers that separate them, conceding that they are "real issues on which clear and honorable thinkers may eternally disagree" (1999:68). I am more optimistic than Hacking on this point, even as I acknowledge, probably more strongly than he, the difficulty of overcoming temperamental barriers. I believe that if one allows oneself in the spirit of a true scholar to survey the array of scientific knowledge available on any subject, the sheer weight of the evidence will eventually wear down ideological and temperamental opposition. There are a number of prominent social scientists who have been dragged by the data to positions they formerly found ideologically distasteful,[2] but let us ignore that point and examine Hacking's three sticking points.

The first sticking point is *contingency*, which is the denial of inevitability, or the notion that nothing in science is predetermined and that it could have developed in many different ways. Constructionists support contingency, and scientists (at least natural scientists) support inevitability. Physicists maintain that any adequate physics would have inevitably evolved much the way it did in any cultural context. That is, the same discoveries, laws, and theories would obtain because there are constraints presented by the hard facts of nature to prevent contingent modifications based on historical or cultural practices. Nature largely dictates the questions of science, and most certainly its answers. Hacking gives Maxwell's equations, the second law of thermodynamics, and the velocity of light as primary examples that any adequately successful physics would have found eventually. This idea of inevitability contra contingency in this sense does not mean that given X, Y is inevitable; it does not lead us back to absolute determinism. Contingency is compatible with probabilistic determinism because it avers that events rely (are contingent on) other prior events to occur. Thus, the truth of a proposition is not guaranteed under any and all conditions. What it means in this context is that had the discoveries Hacking mentions been made in Africa, America, or Asia rather than Europe, they inevitably would have had exactly the same mathematical values; we do not live in an Alice in Wonderland universe.

The knowledge domain of the social sciences, however, is far more open to contingency, because many things in the social world could be otherwise. Indeed, the agenda of left-leaning unconstrained visionaries is to strive for otherwiseness. The extent to which social science concepts contain elements of the natural will be the extent to which the contingency argument is weakened. However, we can rest assured that the inevitability claim will never reach the heights that it has achieved in the natural sciences.

The second sticking point is *nominalism*, which is the denial of abstractions and universal realities. Nominalism questions the correspondence between the names we give to things and the external world. It asserts that the conceptual categories by which we organize the universe do not correspond to any inherent structure of the universe, but rather they are products of the human mind. Scientists adopt a realist philosophy and assert that our nominal categories are natural reflections of real features of a structured universe, and that they depend on the human mind only to the extent that we name, categorize, and order those features.

Nominalism also avers that there is no reality beyond the observable (in this sense, nominalists are radical positivist empiricists). Reality can only exist for nominalists in particulars, never in universals (unseen properties common to the various instances). Humanness, for instance, is instantiated only in particular persons living in the phenomenal world; it does not exist independently in some noumenal world. For something to exist, then, it must occupy space. No one has ever seen humanness, cupness, or redness. Nominalists maintain that these abstractions are just convenient names that help us to classify individual humans, cups, or red objects that we see. Conversely, universalists maintain that to make use of abstract concepts such as humanness, we ignore individual differences such as gender, color, size, age, and so on, and abstract from all instances when we come across what they have in common—that is, their humanity. Extreme nominalists, however, reject even the notion that there is anything common to subjects or objects selected by a name (e.g., human); for them all subjects or objects named are unique representations of the world.

The existence of universals is something that both Plato and Aristotle agreed on, although their concepts about them were radically different. Plato's universals were his archetypal Forms, his perfect humanness, horseness, cupness, redness, treeness, triangleness, and so on, which constitute ultimate reality and which existed prior to any instantiation of these things. In Aristotle's view, universals exist, but only do so (become real) when instantiated. Put otherwise, Plato's humanness existed prior to any human being; Aristotle's humanness is a quality we ascribe to humans, and thus human beings must exist before we can realistically talk about humanness. Aristotle's view is eminently sensible and necessary if we are to talk about the world rationality. Certainly, no one has ever seen abstractions such as strength, intelligence, empathy, or happiness, but we see them demonstrated by humans every day, and they can even be measured, albeit imperfectly. To say that they do not exist as human commonalities differentially experienced and expressed is, in my opinion, pure clever silly folderol.

The third sticking point is the *explanation of the stability of scientific theories.* The constructionist believes that much of the stability of scientific thought is external to the content of science, such as discipline politics, institutionalized knowledge, social and political climate, and funding priorities. Scientists maintain that though these things obviously matter, the ultimate explanation for scientific stability is internal to the nature of science itself. Most constructionists do not deny the reality of scientific facts; they are only insisting that external factors are highly relevant to the stability of scientific knowledge. Nevertheless, there are a radical few who deny any role for the natural world: "The natural world has a small or nonexistent role in the construction of scientific knowledge" (Collins, 1981:3). I think we can safely ignore such an outrageous position, since only the most radical constructionists would subscribe to it.

While it is plain why contingency and nominalism are sticking points, I have difficulty seeing why the stability question is such. The practice of science is plainly a social one in that a number of individuals are engaged in a cooperative process that is governed by rules and enabled or constrained by internal and external events. If China had not shut itself off from the world in the 15th century, we might be now be talking about Wu's equations, Qui's second law of thermodynamics, and Tang's theory of relativity, since China was more culturally and scientifically advanced than Europe at the time. The Cold War greatly accelerated weapons research, AIDS research was necessitated by the emergence of the HIV virus, and only rich countries can afford massive particle accelerator/colliders. It is in this sense, and in this sense only, that we can drag science into historical, cultural, and social relativism. External factors are certainly permissive of science, and thus they dictate much of its practice and progress, but the *products* of science, given adequate funding and a climate of inquiry, depend only on factors internal to it. The claim that the enterprise of science is a product primarily of Western culture is undeniable, but it is illogical to jump from this and say that the products of the scientific enterprise itself—its magnificent theories and all that have issued from them—are without reality independent of culture.

Conclusion

The battle between social constructionism and science is only the latest confrontation in a war that has lasted for millennia. Because it seems bizarre to be against such a magnificent enterprise as science, we might ask if strong social constructionists are really anti-science. Hacking thinks so, and states that "What is true is that many science-haters and know-nothings latch on to constructionism as vindicating their impotent hostility to the sciences. Constructionism provides a voice for that rage against reason. And many constructionists do appear to dislike the practice and content of the sciences" (1999:67). Hacking is not completely dismissive; he ranks himself as a weak-unmasking kind of constructionist (1999:99) who is opposed only to the excesses of constructionism. This is the position I claim for myself.

The real value of social constructionism is that it gives us pause when we start to believe that our social practices are natural and inevitable rather than contingent. However, it is surely not a useful epistemology to guide us in our search for knowledge. Social constructionism is a brake preventing us from going too far in our claims rather than an engine moving us forward. In many ways, constructionism is dangerous when it pushes on the brake too hard and diminishes the importance of science. It has been argued, for example, that schizophrenia is a social construct and thus to search for biochemical causes is futile (Boyle 1990). This is ontological vandalism and epistemological anarchy. I wonder if Boyle thinks we can deconstruct schizophrenia by telling its victims that it is not real and that they should redefine themselves into a non-schizophrenic reality. No amount of time on the shrink's couch ever did that for anyone. Schizophrenia has identifiable causes and can be treated with medications that we would not have had we succumbed to the social constructionist argument. Certainly, mental disabilities of various kinds have been conceptualized differently at different times and in different places, but do we want to go back to the days when "cold mothers" were cruelly blamed for them, or when we stuck people in straightjackets and rubber rooms? These are the only default solutions available to the anti-science crowd if they ignore biological reality.

Endnotes

1.   Science does construct things that previously did not exist that are as natural as Koch's bacillus. These include new animal and crop strains, mice bred specifically to produce cancer cells, new chemical compounds, atom splitting, and even new elements produced in the lab. These things are not magically created; rather, they are creative modifications of existing natural structures.

It is interesting to note that the radical fringe of modern social constructionism has become even more than Latour can stomach. He says that he is "ashamed" that Jean Baudrillard, who wrote some constructionist absurdities about the 9/11 terrorist attacks, is French, like himself (Baudrillard had written of how the attacks were the "happy culmination" of the dreams of people everywhere who hate American hegemony). He also laments that "good American kids" are learning that there is no such thing as natural, unbiased truth and that "dangerous extremists" are using constructionist arguments "to destroy hard-won evidence that could save our lives." He goes on to reluctantly admit that there are real natural facts, but asks, "Why does it burn my tongue to say that global warming is a fact?" (2004:227–228). This amounts to a real about-face for someone who apparently was once a fact constructionist.

2.   Walsh (2009) lists prominent social/behavioral scientists who were previously strict environmentalists and who are now biosocial scientists fully open to integrating their disciplines with the various biological sciences. As far as I know, none of them had an epiphanal experience like Saul on the road to Damascus. All were slowly, and most very reluctantly, dragged by their data to their conversion experience. This can only happen, of course, if one actually believes in data. As Thomas Kuhn warns, those who fail to move with the times will find themselves irrelevant; "retooling is an extravagance reserved for the occasion that demands it," he says, and the wise scientist knows when "the occasion for retooling has arrived" (1970:76). The retooled scientist finds many wonders in the new paradigm:

> Led by a new paradigm, scientists adopt new instruments and look in new places. Even more important, during revolutions scientists see new and different things when looking with familiar instruments in places they have looked before. It is rather as if the professional community has been suddenly transported to another planet where familiar objects are seen in a different light and are joined by unfamiliar ones as well. (1970;111)

Chapter 3

Science and Knowing

Defending Science

To have to defend science is rather like having to defend mother, flag, and apple pie, for science is considered by its proponents to be the apotheosis of the human spirit, humankind's greatest achievement (Atkins 2003). Far from being nature's annihilator, science is the voice by which nature comes to understand herself. Science has lifted humanity to such a level of health, freedom, and comfort undreamt of in the pre-scientific world that it is difficult to see how anyone can think ill of it. Yet, as we saw in the last chapter, there are social constructionists who are openly hostile to both the content and practice of science and who maintain that science is just one way of knowing the world among many equally valid alternatives. Poet/philosopher Anne Carson voices a typical anti-science opinion in her thoughts on progress, facts, and chemistry in a TV program about the Nobel Prize in chemistry (in Herschbach 1996:12–13).

> The Nobel Prize idealizes the notion of progress. My problem is that I don't believe in progress, and I am skeptical of how chemistry is contributing to my humanity. . . . The happy delusion that there are such things as facts . . . underlies the whole progress of science and chemistry. . . . I don't want scientists messing around in the garden of my soul.

A moment's reflection will reveal how chemistry has contributed immensely to the dignity and humanity of anyone who has been ravaged by disease or who might have lost a loved one to accident or illness but for the intervention of modern sciences and the technology it makes possible. Science has provided us with tools to combat the nastiness that raw nature throws at us from cradle to grave. Before the germ theory of disease, anesthetic drugs, and the pharmacopeia of pills that do everything from the deadly serious (killing cancer cells) to the slightly silly (erecting penises), humanity's lot was painful and precarious indeed. Despite some well-known problems that have accompanied its march, science has richly transformed human life beyond the wildest imaginations of our pre-Enlightenment ancestors.

Why do we so often see this rage against reason among ostensibly sensible people? Surely the rage is not aimed at the products of science; after all, I doubt very much that folks like Anne Carson engage the services of witchdoctors and potion brewers to mess around in the "gardens of their souls" when they are sick. Rather, I suspect that the rage is directed at the constraints science puts on what we can claim about the social world.

Social constructionism is so seductive because it puts no such constraints on us since it renders everything relative, and thus it is more generous than science in what it allows us to claim. It lets us make preposterous statements such as, "The validity of theoretical propositions in the sciences is in no way affected by factual evidence" (Gergen 1988:37). If nothing is real and objective, if there are no universal standards with which to judge truth and falsity, we need not produce evidence to affirm or deny a favored or disfavored proposition when a paragraph or two of fuzzy postmodernist prose will suffice to sweep the issue under the rug. If everything is relative, we are relieved of the difficulties of wrestling with the theories and methods of science with its claims of objectivity, and blessed with the freedom to deconstruct concepts we find not to our liking. "Without scientific reasoning as the core value, 'theoretical imagination' is allowed to run amok," assert Wright and Boisvert (2009:1232). How true: it is so much cozier to see what one believes than to believe what one sees. Science upholds positions on "knowing" that strong social constructionists abhor, such as empiricism, objectivity, reductionism, determinism, and some forms of essentialism.

Rationalism

Rationalism and empiricism are rival epistemologies, although neither school of thought disregards the primary tool of the other. The primary tool for rationalism is reason, and for empiricism it is sense experience (Plato vs. Aristotle again).[1] Rationalists regard reason as a uniquely privileged means of acquiring knowledge independent of and superior to experience. For rationalists such as Immanuel Kant, the world comes to us through the buzzing confusion of sense perceptions that must be filtered and organized by the intellect. The world could only be understood through the intellect for Kant, but it is only an understanding of the phenomenal world; the noumenal world of the "really real" is forever closed to us. Rationalists idealize mathematics as the only true paradigm of truth because mathematical thinking is analytic; that is, it rests on a priori knowledge that is true by definition. Deductive top-down reasoning from truths considered self-evident had been taken as the ideal path to knowledge for centuries. Deductive reasoning is ideal because it guarantees the truth of the conclusion given that it is already present in the premise. Analytic statements (e.g., all mothers are females) are broadly tautological in that any denial of them is self-contradictory.

The abstract language of mathematics has been enormously useful in empirical science, with which it is deeply connected, but to expect the tangible world to mirror the perfection of the instrument used to gain insight into it is overly optimistic. As none other than Albert Einstein (a real Platonist who never performed an experiment in his life) put it, "As far as the laws of mathematics refer to reality, they are not certain; and as far as they are certain, they do not refer to reality" (1923:28). But in defense of the connections between mathematics and reality, the heliocentric model of the solar system was mathematical; Copernicus could not directly experience the earth moving around the sun. Einstein's theory of gravity (general relativity) predicting that the light from distant stars would be bent by the sun's gravitational field was mathematical. Observations of this predicted effect could be made only during a solar eclipse, and were made by Arthur Eddington in 1919, three years after Einstein published his theory (Okasha 2002). Modern string theory (the search for the ultimate theory of everything, and understood by almost nobody) is being pursued mathematically and without any experimental guidance (Polsek 2009). It is this kind of thing that led physicist Steven Weinberg to remark that it is "positively spooky how the physicist finds the mathematician has been there before him or her" (in Sarukkai 2005:420). Thus, many discoveries are "rationalized" by the intellect before they are observed by the senses.

Mathematics is enormously useful in science, but some constructionists have tortured the Queen of the Sciences to make some bizarre points that do not map to any kind of reality. Sokal and Brikmont (1998:109) give an example from the work of linguist Luce Irigary, who claimed that Einstein's $E = mc^2$ is a "sexed equation" because it "privileges the speed of light over other speeds that are vitally necessary to us." She never says what the other speeds are or why they may be "vitally necessary to us." If we substituted one of her "other" speeds in the equation, the relationship between mass and energy would no longer hold. For her, Einstein's equation doesn't describe an existing relationship; rather, it created the relationship, and other speeds would have created a different, equally valid, relationship. Irigary also complains about science privileging solid mechanics over fluid mechanics because the latter deals with feminine "fluids," in contrast to the masculine "rigidity" of solid mechanics. Her point is that science is an inherently masculinist political enterprise constructed in the context of a sexist conspiracy cleverly disguised in arcane theories and equations.

I thought that that Irigary's claim was the consummate instance of Charlton's clever silliness until Richard Dawkins (1998) offered us an equally egregious howler from the pen of Jacques Lacan, a man so revered in the postmodernist world because of his pretentious use of mathematics. One would think that this would be off-putting to the anti-science crowd, but since it all turns out to be gibberish, I suppose that's okay. Lacan purported to show that $\sqrt{-1}$ = the human penis!! (I think that warrants two exclamation marks). Perhaps he is demonstrating the sexism of math in that -1 is code for the female absence of a penis (and possibly for Freudian penis envy), or maybe because $-1^2 = 1$, it means that men have "one"; your guess is as good as mine. On the other hand, because $\sqrt{-1}$ is an irrational number, conceivably he is calling the human penis irrational, in which case he may be quite right. It is apparent that one of the aims of the social studies of science as represented by Irigay and Lacan is to torture the theories and equations of mathematics and physics until they confess their malevolent intent. One wonders if fringe constructionist types have followed Alice down the rabbit hole and found their home there puffing on hookahs with the caterpillar. One may also wonder what Lacan's hero Sigmund Freud would have made of all this sexualizing of science and mathematics.

Francis Bacon and Empiricism

On its face, the rationalist ideal of totally reliable knowledge as in mathematics is commendable, but it is an impossible ideal. Francis Bacon, while not the first person to note this, was the first to systematically attack it and defend the empiricist alternative. In his *Novum Organum*, published in 1620, he laid out the biases and prejudices that beset human reason and championed knowledge acquisition that proceeds from induction, which presupposes nothing, rather than from deductive axiomatic truths. He described deep-rooted biases and prejudices that contaminate reason in his famous Four Idols.

The first of these idols is the *Idols of the Tribe*, which have their foundation in human nature and are thus prejudices of reason and of the senses that all humans share. As the Gestalt psychologists point out, the human mind has a great tendency to see more order and regularity in the world than actually exists, a tendency to give substance to abstracts, and a tendency to accept things that we wish to be true and reject those that we wish not to be true. As Bacon put it himself, "the human understanding, from its peculiar nature easily supposes a greater degree of order and equality in things than it really finds" (in Montague 1841:347).

Next, the *Idols of the Cave* are the errors of individuals that exist in addition to the errors all humans share. Bacon attributes the origins of these errors to learning and experience, but surely individual temperaments are of prime importance here.

Then come the *Idols of the Market Place*. Due to the ambiguities of language—words and concepts meaning different things in different contexts—language confuses our understanding of nature. Because of this, different worldviews have a tendency to talk past one another.

Lastly, we have the *Idols of the Theatre*, which are derived from the grand schemes of philosophy to which individuals pledge their allegiance. Bacon tells us that humans have a strong inclination to construct elaborate philosophical systems founded on nothing but our faulty reason and for which there is little evidence from experience.

Because Bacon's *Novum Organum* exposes problems with the tools with which humans construct their opinions and attitudes toward the objects and subjects they perceive—how they create referents in conformity with social interests and desires—it can be loosely viewed as an early work of weak social constructionism. This is particularly evident in his descriptions of the market place and theatre idols, but he went on to uncover what lies beyond the mirages shaped by the errors of sense and reason in the form of empiricism.

Empiricism is the path of modern science. Empiricists do not deny that concepts can be independent of experience but maintain that if those concepts refer to the tangible world, the truth about them can be established only by observation and experiment. Hypotheses are deduced from a form of a priori knowledge that we call theory, but theories are not true by definition; they have been synthesized from numerous empirical facts and are truths that may not survive the tests to which they must be constantly subjected. Aristotle's ideal method of gaining knowledge was deductive reasoning, but he realized that his syllogisms must predispose broad inductions to validate their major premises. Plato's rationalist conclusions become empirical hypotheses in Aristotle's hands. For the empiricist, then, knowledge is a posteriori; we can achieve it only with some degree of confidence after we test our concepts in the world outside our own minds. Empirical statements are thus synthetic, such as "All mothers are nurturing." This is a statement that is not necessarily true, since the predicate is not contained in the subject, and to deny it would not be self-contradictory. It is a statement that can be true or false, and one that we must refer to the stern judge of experience. Empirical science cannot produce the absolute knowledge demanded by those who identify all true knowledge with the magnificent certainty of mathematics.

Bacon and Metaphor

Bacon's frequent use of metaphor left him naked to the barbs of the anti-science crowd within feminism. Many radical feminists claim that Bacon's work is rife with sexist rape metaphors used to persuade us that "the experimental method is a good thing" (Harding 1991:43). Bacon's aggressive empiricism represents the dreaded "master's tools" (Unger 1996) that reveal unwelcome facts (e.g., sex differences) to be challenged. But because the hard facts of science are hard to tackle on their own terms, why not lay bare the rotten core of empiricism instead? Merchant (1980:168) exposes the "hidden agenda" of empiricism by quoting Bacon's advice for relentlessly pursuing knowledge: "For you have but to follow and as it were hound nature in her wanderings, and you will be able when you like to lead and drive her afterward to the same place again. . . . Neither ought a man make scruple of entering and penetrating into those holes and corners, when the inquisition of truth is his sole object."

Read with one's mind ensconced among the idols of the theater, the reference to nature as "her," to the "hounding" of her, and to having no scruples about "penetrating" into "holes and corners," represents Bacon as a barbaric misogynist who modeled his philosophy of science on the rape and torture of women. If one accepts this interpretation, the elision from "rape is bad" to "science is bad" is easily made. It apparently works well, because according to a survey of academic psychologists by Rhoda Unger (1996), most self-identified feminists reject traditional science, viewing it as an outdated patriarchal method of acquiring knowledge that is unsuitable to the feminist enterprise. The very notion of a search for objective truth is considered a male Eurocentric illusion. This strand of criticism seems to be a return to Romanticism without the eloquence of its opposition. For many feminist constructionists, scientists are now not simply emotionless creatures to be pitied, but malevolent oppressors of the poor, of women, and of non-whites, who are to be reviled: "The ideal of the dispassionate investigator is a classist, racist, and especially masculinist myth" (Jagger 1986:158).

There is nothing wrong with using metaphors as heuristic devices. Vickers (2008:125) points out that in his *Rhetoric,* Aristotle recommends their use, "but urged that they 'must fairly correspond to the thing signified: failing this, their inappropriateness will be conspicuous: the want of harmony between two things is emphasized by their being placed side by side.'" Metaphor is the elision from the literal to the figurative, from the underlying principle to the imagery from which we can grasp the author's meaning in the literal. Thus, by transferring the thing being signified to something appropriately signifying that thing, we make mental associations and translate from one domain of meaning to another.

The operative word is *appropriate.* Some anti-science feminists have turned to rape metaphors even where the original works contain no terms such as Bacon's "penetration" and "holes" to indict them. Sandra Harding's (1986:113) description of Isaac Newton's *Principia* as a "rape manual" is an oft-quoted example that misses *appropriate* by a country mile:

> A consistent analysis would lead to the conclusion that understanding nature as women indifferent to or even welcoming rape was equally fundamental to the interpretation of these new questions of nature and inquiry. Presumably, these metaphors, too, had fruitful pragmatic, methodological, and metaphysical consequences for science. In that case, why is it not illuminating and honest to refer to Newton's laws as "Newton's rape manual" as it is to call them "Newton's mechanics"?

I am aware that Harding does not literally mean that the pious Newton wrote an instruction book for sexual predators, but she does consider science as male rape of female nature and further extends her metaphor to marital rape, in which the husband raping his wife is equated with the scientist forcing nature to yield to his wishes. This misandrous mush, combined with her tendency to see ravenous penises lurking everywhere, says far more about Harding's psyche than about Newton's *Principia*. Her work, along with that of Irigary, Lacan, Merchant, Unger, and others quoted here, demonstrates to the skeptic that there are actually folks who disdain science.

Thomas Kuhn and Scientific Relativism and Revolution

Thomas Kuhn's book *The Structure of Scientific Revolutions* (1970) is one of the most influential books in the philosophy of science of the 20th century. Although Kuhn was a strong supporter of science, this book has been used (but not with Kuhn's blessing) to support radical positions about the instability, relativity, and subjectivity of science. Kuhn argued that science does not progress in a neat, linear accumulation of knowledge, rather that science is a messy business that undergoes periodic revolutions that may radically alter the way the content of a particular science is viewed. Kuhn wrote that "normal science" is conducted within a paradigm, which is a set of fundamental assumptions, concepts, values, and practices shared by a scientific discipline that guides its view of reality. Normal science tests hypotheses derived from theories shaped by the contents of the paradigm in which they exist, and to extend the knowledge that the existing paradigm permits, not to look for novelties within it. Work extending beyond what the paradigm permits is rarely tolerated by the guardians of the paradigm, and what will not fit is "often not seen at all" (Kuhn 1970:24). While granting that it is risky to oppose the prevailing paradigm, it is far from true that scientists are imprisoned by it. Many scientists, especially the most open and daring, are attracted to disruptive ideas that challenge the paradigm. New and exciting ideas found outside the box is the stuff out of which Nobel Prizes are fashioned.

As anomalies the paradigm cannot explain accumulate, a crisis ensues, and a new paradigm vies with the old for supremacy. The new paradigm, typically championed by the young and the bold, is engaging in what Kuhn calls revolutionary (as opposed to "normal") science, but when anomalies become the expected, a paradigm shift has occurred.

To give an example from the behavioral sciences, it was a huge shock to socialization researchers to discover the relative unimportance of the shared family environment in producing personality and cognitive similarities among siblings (Rowe 1994). The orthodox socialization paradigm assumed that shared experiences within the family made siblings alike in their psychological development, and that the most important of these experiences was parental treatment. That is, children who are treated affectionately are less antisocial than those who are abused and neglected, and parents who are confident, well-liked, and sociable parents have children who manage their lives well and get along with others. These kinds of results were consistently found, thus confirming that the paradigm worked. This serves as an example of the important Kuhnian concept of "theory-ladenness" of data. That is, data are always contaminated with the assumptions of theory, so even if our data correspond with our theory, it is of little comfort if the data are infected with theories that may be flawed and incomplete.

Then along came researchers schooled in genetics who pointed out that 99% of socialization studies observed only one child per family (Plomin, Asbury, & Dunn 2001). These researchers pointed out that previous findings such as those given above may have had more to do with parent/child genetic similarity than with parental treatment. Because these studies were contaminated by strict environmentalist theory, it simply did not occur to those doing normal science to even consider this. Those working in the revolutionary paradigm asserted that looking at only one child per family does not allow for the assessment of child effects, and thus genetic effects, on parental behavior. Eventually it became clear that to examine the role of these effects, researchers needed twin and adoption studies to tease apart genetic and environmental sources of variation (Grusec & Hastings 2007). Thus, social scientists came to know what every parent who has more than one child knows—there are different parenting styles for different children. The same parent who is permissive with a warm and compliant child may be authoritarian with a bad-tempered and resistant child, while all the time trying to be the authoritative parent that child psychologists tell us that all parents should be to all offspring.

Kuhn claimed that because the terminology and conceptual framework of rival paradigms don't mesh, they are incommensurable, and thus they cannot talk to each other. This was the case for many years in the social/behavioral sciences, as "genetic determinists" and "environmental determinists" vehemently argued the now (hopefully!) moribund nature vs. nurture issue. Because rival theories are incommensurable, it was supposed that one cannot make a rational choice as to which is superior, a position that seems to entail that theory choice in science is irrational—in other words, determined by factors external to science, and thus relativistic.

This is the claim that radical constructionists have jumped on to discredit science. But they did not look closely before they leaped. Kuhn was not asserting that science is subjective and irrational, with its theory choices being made only by mob consensus. It only seems that we cannot rationally choose between theories if we are in the middle of the paradigmatic crisis and are stubbornly ignorant of the claims put forth by the proponents of the revolutionary paradigm. Kuhn points out that a hypothetical observer ignorant of the chronology of events who was asked to choose the superior theory would do so every time, and that it would be the most recent theory because it would be the one best able to accommodate the known facts of a particular domain. Kuhn concluded that the preceding point "is not a relativist's position, and it displays the sense in which I am a convinced believer in scientific progress" (1970:206). In a later (1977:35) work he even listed the qualities of a good scientific theory: it should comport with facts derived from experimentation and observation, be logically consistent in that it fits the known facts and principles of the domain together harmoniously, be broad in scope and simple (Occam's razor); and be forward looking (leading to the discovery of new facts) as well as backward looking (systematizing the known facts).

Kuhn seemed to have been an accidental radical—and certainly a false prophet to those who wish to denigrate science. Of course, science is an unstable, messy, theory-laden social enterprise, but so what? If a scientific theory is transient, it is because the scientific process has found a theory that better comports with reality. The tempo of science advances much like Gould and Eldredge's (1977) punctuated equilibrium model of evolutionary change, which asserts that evolution involves the slow accretion of adaptations and long periods of stasis that are punctuated periodically by very rapid changes. Both the pace of biological evolution and science is accelerated by environmental events—in the first case, by something like climate change or the introduction of a new predator, or in the later by the introduction of a new instrument (telescope, microscope) or the introduction of concepts from adjacent sciences, such as the introduction of atomism into chemistry or genetics into the social and behavioral science. As Fromm (2006:583) points out, scientific disciplines are in "constant flux from right, to more right, to even more right. If they reach stasis they're dead." We reach the truths of our disciplines tentatively and asymptotically, and only an incurable dreamer would believe that we will ever touch the axis of ultimate *Truth*. Science thrives on ignorance, as Matt Ridley (1999:271) tells us: "A true scientist is bored by knowledge; it is the assault on ignorance that motivates him - the mysteries that previous discoveries have revealed. The forest is more interesting than the clearing"

Relativism

The idea of relativism is trivially true: Is Omaha, Nebraska, east or west; is polygyny illegal; is the pope, speaking *ex cathredra*, infallible? The answers to these and thousands of other similar questions depend on where you are, who you are, and what you believe. The central ideas of relativism are that either there are no absolutes and no objective epistemic standards by which we can judge anything true or false, or that there are many such standards, each true within their own contexts. Many of us might agree with either or both of these positions as they apply to the vast majority of moral, esthetic, religious, historical, cultural, and political issues, but radical social constructionists have gone further to make the same claims about science.

Now, it is certainly a good thing to be skeptical about claims of truth, even scientific claims, for the credo of science is question everything and believe nothing until it is proven beyond a reasonable doubt. My problem is not necessarily with the varieties of philosophical relativism, but with the appropriation of crude forms of it by social constructionists to cast unreasonable doubts on scientific claims for which there is an abundance of hard evidence in order to advance an ideological agenda.

In his *Fear of Knowledge* (2006), Boghossian claims in that relativism is used by constructionists as a shield to protect them from their "fear of knowledge." Boghossian wants to convince relativists to examine their motives for hanging on to relativistic beliefs, which he sees as an incoherent attempt to appear as open minded and non judgmental liberals. The relativist argument is that in the absence of any absolute standards for deciding among conflicting beliefs of right and wrong, all cultural value systems are equally valid. This limits discussion of issues of morality and truth to descriptive and non-normative discourse, amounting to intellectual laziness because it leads to the conclusion that we can rest content with "truth" being whatever happens to be true for us. Those who criticize the Nazi holocaust, racist and sexist practices, the execution of homosexuals, cannibalistic practices, and any number of other obnoxious practices have no defensible grounds to do so from a relativist position (Boghossian 2006).

The relativist argument extends beyond moral issues of right and wrong to scientific questions of truth and falsity. Boghossian relates a story about the Lakota tribe of South Dakota, who believe that they are descendants of Buffalo people who emerged from a subterranean world to the surface to prepare it for humans (2006). This is a story as culturally valid as any number of creation myths from around the world, and Boghossian's quarrel is not with mythology or with the Lakota. He is appalled, however, with the relativism of many academic anthropologists he quotes as defending the myth as an equally valid way of knowing about the origins of humanity. Apparently, the academics he quotes as defending this and other creation myths (the Christian version excepted, of course) as co-equal with Darwinism would have no qualms about them being taught as equally valid accounts of the origins of humans in biology classes, since science is only one of the many ways of knowing the world.[2]

I am fairly sure (and I'm fairly sure that Boghossian is fairly sure) that none of the anthropologists who made the "equally valid" statements in public (in the *New York Times*, in this case) really believe them. The position that no standard of knowing is privileged over others is the sort of thing that gives relativism a bad name. Even Richard Rorty, an arch relativist, dismisses the position as "silly" (1991:89). Such extreme relativism renders it impossible to believe that anyone, including ourselves, can ever be in error and demolishes the distinction between what we choose to believe and what is actually true. But then, the lack of vigorous standards by which to judge truth and error is a big part of the reason that many well-meaning social scientists gravitate to social constructionism.

With Boghossian, Mark Kalderon (2009:238) insists that the source of strong relativistic conviction cannot be in the cogency of its arguments since they are, in the main, incoherent and self-refuting. A self-refuting argument is one that logically contradicts itself by holding itself to be true. A crude example is: If *all* truth is relative, then the claim that all truth is relative is absolutely true. If it is absolutely true, it is obvious that not everything is relative, rendering false the statement that "all truth is relative." Plato exposed the self-refuting nature of strong relativism (interestingly, a position held mostly by the ancient sophists whose system of thought—sophism—is now used as a synonym for specious or deceptive) long ago in the *Theatetus*, as did Aristotle in his *Metaphysics*.

Kalderone agrees with Boghossian that there is a genuine fear of knowledge in the constructivist camp because "relativist conviction is animated by the thought that the authority of reason, and its attendant rhetoric of objectivity, is a mask for the interests of power." He goes on to assert that any effective case against relativism must deal with this fear, because reason and objectivity are sometimes misused in the interests of power or else risk "further entrenching the relativist conviction" (2009:239). Any system of thought that actively seeks to prevent the march of science (as opposed to monitoring it and demanding practical and ethical justification) because it finds things in it that it fears is a system that represents the betrayal of the Enlightenment tradition of human progress. For my part, I am more inclined to the opinion that fear of knowledge can be quieted by actually learning about what is feared. Fear is fed by ignorance, which is easily rectified if one is willing to spend the time and effort, because as previously noted, science is also fed by ignorance, albeit an ignorance that motivates a desire to know. Those who do so may find that there is nothing to fear from truth.

Conclusion

Science is the greatest intellectual achievement of humankind and arguably the only source of reliable knowledge about anything natural in the universe, and it has enriched human life immeasurably. Science is superior to other ways of knowing because it yields justified beliefs—that is, beliefs that are verifiably true across all cultures—and thus produces better explanations of reality as determined by how it systematizes and unifies knowledge in a domain of inquiry. It is certainly not the only way of knowing, and it may be inferior in some ways to more emotionally satisfying ways of knowing. Religion, poetry, art, music, literature, and philosophy may better satisfy the human spirit, although many see deep emotional beauty, awe, and wonder in science too. Science does not claim to answer what may be the most meaningful of personal questions: Who am I? What is the purpose of life? What is my ultimate fate? These questions may be answered to different degrees of satisfaction by different philosophies, but the answers do not translate from person to person and from culture to culture in the same way that scientific answers (albeit to less personally profound questions) do. Questions posed about the tangible world, the world we seek to understand and make better, can only be addressed reliably by the tried and true methods of science. Science doesn't always get it right—and sometimes even gets it horribly wrong. But scientists know that their work is always tentative and self-correcting should it be in error. Relativists are relieved of the burden of being in error, since for them there is no objective way of determining truth and error. Either there is no truth or there is a plurality of truths, all correct.

Endnotes

1. Plato's Academy was devoted to analytical thought, as the sign above the Academy entrance made clear: "Let no one ignorant of geometry enter here." Aristotle founded his own school (the Lyceum) after Plato died emphasizing the substance of physical things, the knowledge of which came from the senses. His work on astronomy, physics, and biology, although seriously flawed, was the final word for almost two thousand years.

2. Anthropologists have long debated whether the knowledge they present to the world is objective or subjective, and they introduced the terms *emic* and *etic* as a solution around that philosophical problem. An emic account is an account of some phenomenon provided by members of cultures being studied, such as the Lakota's account of creation. An etic account is an account of the same phenomenon by an outsider, such as an anthropologist. It is bad practice to confuse these two types of accounts when the phenomenon being described is of concern to science rather than cultural practices, beliefs, or morality, as in this case.

Chapter 4

The Conceptual Tools of Science

Terms of Opprobrium

Every discipline has its ontology, which informs it of the fundamental categories of reality within its domain, and its epistemology, which defines how its practitioners know and reason about that reality. Ontology is the biologist telling the philosopher that this animal specimen is female; epistemology is the philosopher asking how the biologist knows that. The biologist will then explain the methods by which sex is categorized in biology and describe the various things that differentiate one sex from the other. He or she will pay particular attention to distinctions that are necessary to placing a specimen into one category versus the other, such as chromosomal and gonadal status. To do this, the biologist will have to admit that he or she employs methods and concepts that are anathema to the social constructionist wing of social science. These interrelated methods/concepts are determinism, reductionism, and essentialism. The biologist's explanation is reductionist because it reduces this female to her chromosomes and ovaries; it is essentialist because these are the essential factors of sex differentiation, and it is deterministic because these factors are said to determine or cause this specimen to be female.

As with other isms, determinism, essentialism, and reductionism have gradations of acceptance, but opponents tend to attack only the most extreme versions, as if they were their sole representations. I have attacked extreme versions of constructionism and relativism, but I have acknowledged the usefulness of the more moderate versions. Few social constructionists will likewise recognize the usefulness of non-extreme versions of determinism, reductionism, and essentialism, since those who use them are considered not only wrong, but malevolently wrong.

Determinism and Freedom

Characterizing determinism as "malestream," Bilal Shah (2009:1) writes that "Determinism and feminism are as antithetical to each other as sugar is to spice, fire to water, and day to night. Determinism implies that human action is determined by forces independent of will. Feminism implies that structuring of production, reproduction, sexuality and socialization which have put women at a disadvantage can be deconstructed bycollective will." Shah is making a number of explicit and implicit claims here. The first is that determinism is owned by males, which is nonsense since it is simply a concept about how the world works held by scientists, male or female. Second, she talks of human will as if it is separate from nature—a ghost in the machine, perhaps? Third, she is herself deterministic when she assumes that people can cause a change in the way the world is by collective will. She is saying that if we do XYZ, we should be able to affect a change in P, assuming the absence of potential disruptors. What is that but determinism? Fourth, she assumes that determinism means that humans cannot change the course of future events. To assert this is fatalism, not determinism, a position it would seem that Shah and many others like her apparently confuse with determinism.

Enemies of determinism equate determinism with universal determinism, a view usually associated with 18th-century French mathematician and astronomer Pierre Laplace, who wrote:

> An intelligence knowing all the forces acting in nature at a given instant, as well as the momentary positions of all things in the universe, would be able to comprehend in one single formula the motions of the largest bodies as well as the lightest atoms in the world, provided that its intellect were sufficiently powerful to subject all data to analysis; to it nothing would be uncertain, the future as well as the past would be present to its eyes. The perfection that the human mind has been able to give to astronomy affords but a feeble outline of such an intelligence. (in Bishop 2006:2)

This certainly looks a lot like fatalism, or perhaps Calvinistic predestination, since only an almighty God could possess such an intellect. But Laplace was well aware of this and posed this hypothetical as true in *principle* but impossible in practice. He posed this impossible scenario precisely to propose how science should proceed in the absence of perfect knowledge—that is, to contrast the concepts of certainty and probability and to introduce his statistical physics (Bishop 2006). Yet we still see instances when scientists hold Laplace's purposely constructed straw man as the true exemplar of determinism (Saint-Amand 1997).

When scientists speak of determinism, they usually mean causal determinism, which simply means that every event stands in some causal relationship to other events. Determinism is a position relating to how the world is said to operate and is a position held by all scientists, male or female; what would it be for a scientist *not* to be a determinist? Scientific determinism does not state that X will lead to Y absolutely and unerringly; rather, it says that given the presence of X, there is a certain probability that Y will occur—which was exactly Laplace's conception of determinism. Surely we are all determinists in this sense. Certainly, there are unique events and random happenings (thus we cannot always speak of statistical regularities, although even unique events and random happenings have causes), but the world is not chaos and randomness; it has a great degree of predictability about it.

It may even be that determinism is necessary for freedom and agency. If I did not think that the things I do produce (determine) meaningful consequences, why would I do anything? All rational action and education is deterministic. Free will/agency and determinism can conceptually and peacefully coexist as compatibilists propose. I know that I am a free agent and that living according to that position is necessary, but I also know that my agency is constrained and/or enabled by my temperament, upbringing, knowledge, conscience, and physical and cognitive abilities and disabilities, as well as the constraints imposed by others. If my agency extends only to following the strongest inclination congenial to my nature, this is not freedom for some people. But I have no qualms with this kind of freedom; after all, my strongest inclination is *my* strongest inclination, and no one else's. My nature is me and my will is mine, so if I follow the direction my nature nudges me in, I am following my will. To ask for freedom beyond this makes no sense, for how can one be free of one's nature?

Yet it is popular in some new-age quarters to celebrate quantum mechanics in the form of Heisenberg's principle of uncertainty as destroying determinism and affirming the unfettered freedom of the will. But who wants the kind of freedom in which no one can probabilistically predict the behavior of anyone else? Max Weber wrote of this kind of freedom as the "privilege of the insane" (in Eliaeson 2002:35).

Determinism gives us the only kind of free will worth having. Writing of the uncertainty principle, Nobel Prize–winning physicist Richard Feynman says that the prediction of an emission of a photon from an atom is probabilistic but still deterministic, and that "This has given rise to all kinds of nonsense and questions of freedom of will, and the idea that the world is uncertain" (in Corredoira 2009:450). The indeterminacy of quantum phenomena is not in the system itself but exists "only when the measurement is carried out" (Corredoira 2009:450). There are no uncaused causes miraculously free of nature.

"Biological Determinism"

When social scientists use the term determinism they are typically thinking of biological determinism; cultural determinism is apparently acceptable. Biological determinism is seen as implying that social behavior is a direct outcome of genetic programming absent any influence from the environment. Colin Trudge (1999:96) opines that such reasoning represents either mere rhetoric or simple ignorance: "For a start, no evolutionary psychologist [or geneticist or neuroscientist] doubts that a gene is in constant dialogue with its surroundings, which include the other genes in the genome, the rest of the organism, and the world at large." If only those who make charges of "biological determinism" would learn something about human biology, they would not embarrass themselves by reflexively bandying about that naïve accusation.

An additional concern for many social scientists is that explanations of human behavior are socially dangerous if there is even a whiff of biology attached to them. Notwithstanding the fact that there is no such animal as a strictly biological explanation for social behavior, far more damage has been done by a blank slate view of human nature, which implies a more sinister form of determinism. Stalin, Mao, Pol Pot, and others of similar mindset murdered in excess of 100 million people in their belief that they could take empty organisms and turn them into the "new Soviet, Chinese, or Cambodian man" (van den Berghe 1990:179). A blank slate view of human nature is the delight of political megalomaniacs who believe that humans can be molded into whatever conforms to their vision of social perfection. A view of human nature that sees each person as a unique individual born with a suite of biological traits that will influence the way they will interact with the world is more scientifically defensible and respectful of human dignity than tabula rasa views based on Utopian dreams.

Matt Ridley (2003:6; emphasis added) had something quite astute to say about the fears of genetic determinism and why social scientist should rid themselves of such fears:

> Genes are not puppet masters, nor blueprints. They may direct the construction of the body and brain in the womb, but they set about dismantling and rebuilding what they have made almost at once in response to experience. They are both the cause and consequence of our actions. Somehow the adherents of the "nurture" side of the argument have scared themselves silly at the power and inevitability of genes, and missed the greatest lesson of all: *the genes are on our side.*

Ridley is saying that genes are at our beck and call, not we at theirs. Genes are constantly responding to our needs by making the hormones, neurotransmitter, and cell-structure proteins we need as we meet the many challenges of our environments. Badcock (2000:71) goes so far as to assert that our genes "positively guarantee" human freedom and agency. If they incline us in one direction rather than another, we are being nudged internally, not by something wholly outside of our beings; after all, our genes are *our* genes. Likewise, because so many things that we do in life affect the expression of our genes, epigeneticist Randy Jirtle asserts that "Epigenetics introduces the concept of free will into our idea of genetics" (in Watters 2006:34).[1] And finally, neurobiologist Steven Rose writes, "Individually and collectively we have the ability to construct our own futures, albeit in circumstances not of our own choosing. Thus it is that our biology makes us free" (2001:6). Very few social scientists would feel as comfortable making such bold pronouncements about free will and agency.

Essentialism

Essentialism is a concept caught up in such arcane metaphysics that neither defenders nor opponents seem quite sure what they are defending or attacking. For Plato, an "essence" tied to any object, subject, or substance is integral to his permanent, unalterable, and eternal Forms. Psychological essentialism appears to mean that human beings are predisposed to see things in terms of "natural kinds" that contain essences defining their nature. Marxist essentialism lies in his concept of the human "species being" in contrast to other animals. The following discussion is based not on any of these versions but on Aristotelian essentialism.

Wood and Eagly (2002:700) define essentialism using sex differences as an example: "Essentialist perspectives emphasize the basic, stable sex differences that arise from causes that are inherent in the human species such as biologically-based evolved psychological dispositions." Few scientists would disagree with this, but most social constructionists certainly would. What constructionists tend to have in mind is the Aristotelian notion that things have essences that are necessary, unalterable, and indispensable to them. An essential property is a property of an object or subject which, if lacking that property, the object or subject cannot be what it is alleged to be. Such properties are universal in every entity that belongs to a particular classification and are not idiosyncratic or context dependent. This claim does not mean that essentialism homogenizes all subjects classified into a group, as constructionist charge. Aristotle distinguished *per se* and *per accidens* properties of an entity. A *per se* property is an essential property that a subject or object possesses and cannot lose without changing its nature; a *per accidens* property is a non-essential property that a subject or object may lack, or lose if it previously had it, without changing its nature.

For instance, it is essential that a molecule of water have two hydrogen atoms and one oxygen atom; without either we do not have a molecule of water. These elements are *per se* properties of water; its manifestations as a liquid, solid, or gas are *per accidens* properties made possible by its *per se* molecular structure interacting with the ambient temperature. Likewise, we can say that it is essential to being male that a person has testes and an XY karyotype, and that it is *per se* essential to the being female that a person has an XX karyotype and ovaries. The constructionist argument is that we cannot say that chromosomal or gonadal status is always and unequivocally associated with gender (or even sex, since some individuals have their gonads removed for medical reasons) in the same way that the combination of hydrogen and oxygen are to water. What counts as an essential property need not be necessary and sufficient for defining a kind, only necessary. A person can be sexed a male yet gendered a female *per accidens*, but it is surely not a sin to define sex according to chromosomal and gonadal status, or even to say it is essential to do so, since it is empirically true in all but the rarest of circumstances.

Andrew Sayer (1997:462), a philosopher/sociologist who believes that "moderate essentialism" is necessary for explanation and critical for social science, notes that critics not only assert that essentialism involves essences, but that these essences are "unchanging eternal ones." He goes on to say that "This helps to load the dice against essentialism." As the *per se/per accidens* distinction clearly highlights, essentialism does not claim that all members of a class of people are identical, only that they share the same essential properties as "natural kinds." They can be radically different in multiple ways in their *per accidens* properties while sharing what makes them what they are and distinct from the things that they are not. As Odeberger (2011:89) says about definitions of natural kinds, "That definitions range from partly accurate [e.g., man is a rational animal] to completely accurate [e.g., silver is a transition metal with element number 47] does not militate against the fact that achieving any degree of accuracy in definitions requires attending to characteristics of the object to be defined."

Before Darwin, biological species were almost universally considered eternal and unchanging, but now we know that even "natural kinds" change over evolutionary time by gaining and losing properties; thus, even essences used to differentiate species cannot be said to be eternal. For Aristotle, the major differential between humans and other animals was reason, because "reason is the master regulatory natural function by which individuals enter into social life" (Wakefied 2000:17). Yet the reasoning capacity of our species has changed enormously from early hominid species to *Homo sapiens*, as indicated by huge increases in cranial capacity. From the approximate 1.5 million years that separated *Australopithecus afarenis* and *Homoerectus,* hominid cranial capacity doubled from a mean of 450 cc to a mean of 900 cc, and by another 70% to about 1350 cc, from *Homoerectus* to modern *Homo sapiens* (Bromage 1987).

Enemies of essentialism often feel so strongly that they attack generalities as essentialist. Janis Bohan (1993:7) provides an example of her vision of essentialism when she states that "If 'friendly' were gendered, an essentialist position might argue that women are more friendly than men . . . and the quality is now a trait of women." She goes on to assert that this kind of generalizing is grounded on "problematic universalizing assumptions" that portray women "as a homogeneous class" that fails "to acknowledge diversity among us" (1993:8). In their battle against false generalities, anti-essentialist feminists have dismissed all generalities to affirm the existence of nothing but context-specific differences in a sort of extreme nominalism. Surely we all recognize that when we make general statements it is a given that there are individual exceptions to the rule without having to make it explicit. Anyone who has taken statistics knows that when we compare mean differences across groups, there is always variance within the groups being compared as well as between them. Recognizing general features of things does not commit us to the doctrine of essentialism as defined by its detractors.

If we define essentialism as the process of generalizing, we are rejecting nomothetic science in favor of idiographicaccounts of contingent and subjective phenomena. This is acceptable to social constructionists who treasure subjectivity and relativism, but it is a major concern for the science-friendly wing of social science. Idiographic accounts have their place, but that is on the therapist's couch or the biographer's word processor, where the focus is rightly on the individual. While it is true that each individual has a unique psychological structure, all science tells us that n = 1 is bogus for advancing our knowledge of human nature and human behavior. We cannot avoid categorization if we are to make sense of the world. Categorization is the search for common properties, which presupposes diversity rather than the homogeneity of that which is being sorted. The features that the category holds to be essential to being placed in a category is a matter to be determined empirically, not by ontological fiats.

I believe that for the most part the charge of essentialism is so much metaphysical waffle. Aristotelian essentialism means "invariant" only in those things that are truly necessary to something being one thing and not another, and surely not even the most radical constructionist believes that there is anyone who thinks that anything in the social world is invariant. But the essentialist term is useful to reinforce other hissing suffixes such as sexist, fascist, racist, or classist, so beloved by the politically correct to stifle inquiry and to congratulate themselves for not belonging to any of those nasty "essentialist" categories.

Reductionism

Determinism and essentialism are terms of opprobrium used mostly by strong social constructionists, but reductionism tends to be used pejoratively by most social scientists. Owen (2006:900) defines reductionism as "attempts to explain social 'reality' in terms of a single, unifying principle such as 'patriarchy.'" If this is what social scientists typically mean by reductionism, I too would call it nasty names, because I snort with impatience when I hear colleagues complain that all the woes of women can be traced to patriarchy or that all the woes of everyone are the result of capitalism. In a perverse kind of way, such thinking is highly deterministic and essentialist as well as reductionist. The reductionism that most social scientists abhor is not this, however, but rather the process of taking causal explanations from higher, fuzzier levels, to deeper, more precise levels. Although reductionism is nothing more sinister than this, some social scientists recoil in horror, as if it would foreclose on their whole enterprise if they stooped to reducing Durkheimian "social facts" to something more elementary. Reductionists do not suggest that the social factist paradigm is wrong, and they appreciate that while social facts do not occupy space, like gravity their effects are real and are revealed in the enabling and constraining effects they have on human action.

James Coleman offers an anti-reductionist example with his assertion that when two or more individuals interact, "the essential requirement is that the explanatory focus be on the system as a unit, not on the individuals or other components which make it up" (in Wilson 1998:187). Coleman declares that social facts *must not* be reduced to individual psychology or biology, because a separate reality emerges when parts belong to a dynamic system such as a human group. It is true that the interaction of elements (whether they be chemicals, neurons, genes, people, or whatever) produce effects that can be explained on their own terms, the claim that it is *essential* to focus explanatory efforts only on the whole unit to the exclusion of the parts is unnecessarily constraining. E. O. Wilson (1998:187) pointed out in response to Coleman that biology "would have remained stuck around 1850 with such a flat perspective" if it had taken seriously the claim that "essential requirement is that the explanatory focus be on the organism as a unit, not on the cell or molecules which make it up."

As Wilson implies, cell biologists know that at bottom they are dealing with atomic particles and seek to understand their properties. But as Coleman might stress, they also know that there are properties of the cell that cannot be deduced from those particles *a priori*, that they require functional explanations of the whole cell and how that cell fits into a network of other cells to form the organism. Both men are right but need to add that we need both holistic and reductionist accounts that complement one another rather than exclude. Science is eclectic by nature and can pose questions and offer explanation at several levels of understanding. Natural scientists have long recognized the complementarity of reductionist and holistic explanations. Useful observations, hypotheses, and theories now go in both reductionist and emergent directions in sciences: from quarks to the cosmos in physics and from nucleotides to ecological systems in biology.

Philosopher of science Thomas Nagelhas pointed out that "non-reductionist accounts simply *describe* phenomena while reductionist accounts *explain* them" (in Rose 1999:915). To explain something, we have to discover its causes, and to do thiswe have to look at its constituent parts. Reductionism and determinism are thus joined at the hip, because the reductionist goal of explanation is intimately tied to the determinist goal of prediction. The correctness and utility of any explanation (whether or not we judge it reductionist) can only be gauged by its predictive power. Social science explanations of broad categories of people such as classes, genders, and races are really descriptions that beg a multitude of questions rather than explanations. Not digging below the surface and ceasing the search for explanations with social facts may be true to Durkheim's dictum, but it is poor science. I am particularly fond of Steven Pinker's (2002:72) account of the differences between reductionist and non-reductionist explanation as "the difference between stamp collecting and detective work, between slinging around jargon and offering insight, between saying something just is and explaining why it had to be that way as opposed to some other way."

Assuredly, there are times when non-reductionist accounts are more coherent and satisfying than reductionist ones. We must be careful that we do not lose meaning as an essential component to understanding behavior by an overemphasis on mechanisms. Phenomena may be explained by lower-level mechanisms, but they find their significance in more holistic regions. We can agree with the Romantics that propositions about entities such as genes, hormones, and neurons do not contain terms that define the most meaningful aspects of the human condition such as love, justice, morality, and awe of the beautiful.

Neuroscientists, for example, have found the neurochemical basis for romantic love in a chemical soup of serotonin, dopamine, and norepinephrine sparking up the nucleus accumbens (the brain's major pleasure center) in brain scans of people in love (Esch & Stefano 2005; Fisher, Aron, & Brown 2005). I wonder what Wordsworth would have said about reducing love to the soup and sparks of brain activity?

Whatever he would have said, I would reply that this in no way reduces the wonder of love as it is experienced, nor does it come remotely close to explaining why a particular Jack fell in love with a particular Jill. All it does is explain what happened in his brain when he did. If I may steal shamelessly from Darwin, "I see grandeur in this view" because it brings us closer "to understanding the neural basis of one of the most formidable instruments of evolution, which makes procreation of the species and its maintenance a deeply rewarding and pleasurable experience, and therefore ensures its survival and perpetuation" (Bartels & Zeki 2004:1164). I think Wordsworth would have eventually shared this view after recovering from the initial shock.

This real concern of social scientists about reductionism is that explanation at lower levels entails the elimination of higher-level explanations. Eliminative reductionism means that higher-level phenomena can be fully explained by lower-level properties, and the higher-level properties have no causal impact independent of their parts. For instance, someone might claim that social class has no impact on a person's behavior once we consider the traits of that person that are assumed to explain both his or her social class and the behavior in question. Although lower-level variables may explain away the direct effects of a higher-level variable, the higher-level variable's indirect effects are not undermined. This is so in the present example because we cannot know all of the causes of each person's social class, and surely the effects of social class amount to more than the sum of its parts. Perhaps all holistic explanations in the social/behavioral sciences are ripe candidates for causal reductionism, but I cannot think of one that is a candidate for eliminative reductionism. We cannot dissolve social and psychological reality into biological processes, but we cannot deny that these processes help to elucidate them. Neither must we confuse a part, however well we understand it, for the whole.

Eliminative reductionism is what Daniel Dennett (1995:82) calls "bad" or "greedy" reductionism, which entails skipping over several layers of higher complexity in a rush to fasten everything securely to a supposedly solid foundation. He likewise describes an antireductionist as someone who "yearns for skyhooks," a sort of *deus ex machina* that can miraculously lift them out of scientific difficulty (1995:82). Dennett's "good reductionism" is simply "the commitment to non-question-begging science." Bad reductionism can be utterly incoherent. For instance, I know that when the aggregate of the water molecules gyrating in my kettle reach 100 degrees Celsius I shall have my cup of tea, but I would lose coherence were I to try to translate this information about temperature, which is an emergent function of the agitation of aggregated molecules, into the motion of individual molecules. Reducing temperature to kinetic energy is an explanation of temperature, but it is not very useful in the kitchen.

Since I have broached the molecular with the above example, it is interesting to see what philosopher of science Nancy Cartwright has to say about reducing chemical phenomena to quantum mechanics without remainder: "Quantum mechanics is important for explaining aspects of chemical phenomena, but always quantum concepts are used alongside sui generis—that is, unreduced concepts from other fields. They do not explain the phenomena on their own." (1997:163). This is exactly the position of social and behavioral scientists who wish integrate fundamental biological concepts into their fields; in other words, they are important but "do not explain the phenomena on their own." Gene x environment interaction models favored by biosocial scientists assert that genes have this or that effect only when certain environmental conditions are satisfied. To examine only genes or only environments when trying to explain human social behavior is to miss half of the picture, and is a fool's errand. This is why we will always need the social/behavioral sciences and their concern with emergent meaning. Nevertheless, when mechanisms are discovered and understood, we can more fully understand and appreciate the emergent phenomena they underlie. Science has made its greatest strides when it has picked apart wholes to examine the parts and in doing so has gained a better understanding of the wholes they consti-tute. As Matt Ridley (2003:163) opines, "Reductionism takes nothing from the whole; it adds new layers of wonder to the experience." We shall know that the social sciences have matured when accusations of reductionism are consistently met with Dennett's mocking answer: "That's such a quaint, old fashioned complaint! What on earth did you have in mind?" (1995:81).

Conclusion

Let us concede that it is true that traits and behaviors imputed to women and racial minorities and considered biologically (and thus dis-missed as deterministic, essentialist, and reductionist) have been used to oppress them, and many believe that modern biology can still be used for the same purpose. Biological findings can be used by misogynists and racists to denigrate and oppress only if we allow them to, and only if we do not counter the ignorance that underpins their arguments. Such people will climb aboard any vehicle that takes them where their prejudices want to go. The pursuit of social justice is a moral imperative, regardless of what science does or does not have to say about any observed differences between the sexes and races. Science must be our unfettered guide to understanding human behavior because it is about what empirically *is*, not what morally *ought* to be. Justice does not rest on sameness or on differences between the sexes and races, but on law and reasoned discourse. Hacking (1999:96) points out that feminists often see the tools of science as tools that "have been used against them . . . women are subjective, men are objective. They argue that those very values, and the word objectivity, are a gigantic confidence trick." Whatever legitimate complaints women (or any other group with discriminatory complaints) have about their treatment, it is hardly legitimate to say that because these tools have been used in this way, the tools themselves are illegitimate. Believing that it is legitimate to blame to tools can only lead to a fear and loathing of science, and that is hardly a good thing in academia, and it erodes public support for education. I would certainly not spend my hard-earned money on a college education for my children if I knew that they were going to hear from faculty in the more intellectually lax disciplines that there are no knowable truths and, worse yet, no tools available to find them if there were.

Endnote

1.   Epigenetics entails any process that changes gene activity without changing the DNA sequence, and it puts a whole new face on the meaning of gene-environment interaction. Epigenetic modifications affect the ability of the DNA code to be read and translated into proteins, making the code accessible or inaccessible and thus influencing the reaction range of a gene. DNA itself only specifies for transcription into messenger RNA, which itself has to be translated by transfer RNA and assembled by ribosomal RNA. Genes are switched on and off by signals from the organism's internal chemical environment and/or by its external physical and social environment according to the challenges it faces. It is in this sense that what we experience and what we choose to do in life influence gene expression. Altering gene activity, of course, alters phenotypic traits. The field of epigenetics shows why a person's genotype does not map unerringly to a particular phenotype (a phenotype is anything—blood group, height, IQ, conscientiousness, and so on—that is observable and measurable), but rather to a range of phenotypes. The genome is not the architect's blueprint; it is the housewife's recipe. Like a fine cake, the form our various phenotypes will take relies

not only on the variety, quality, and quantity of the ingredients, but also on variations in oven temperature and length of baking. The environment thus affects our genes just as surely as our genes affect the environments we make for ourselves. A fascinating introduction to this area understandable to those with a moderate acquaintance with genetics is the journal article "Behavioral Epigenetics" by Barry Lester and his colleges (2011).

Chapter 5

Human Nature

Is There a Human Nature, and How Do We Know?

One of the most fundamental questions we can ask about ourselves is "What is human nature?" Let us acknowledge that the concept of human nature is not sharply defined. Ever since Darwin we have come to expect fuzzy boundaries around the taxonomies of biology. Fuzzy or not, all theories of human conduct contain an underlying vision of human nature, although these visions differ radically, and some claim that there is no such thing. The path of least resistance for a number of social scientists has been to claim exactly that, thus relieving them of the burden of pondering it. A claim that humans have a nature is a claim from weak versions of realism and essentialism as opposed to nominalism and social constructionism. It is a realist position, because it appeals to the reality of the abstraction called humanness and to a universalism asserting that every member of the species *Homo sapiens sapiens* shares this abstraction. It is essentialist by the same criteria; that is, every member of the species has some qualities essential to being included in that species rather than in some other. Although the claim that humans have a nature is realist and essentialist, it is not an assertion denying weak versions of nominalism or social constructionism. It is not anti-nominalist in that it does not deny that each instantiation of that humanness is a unique representation of it, and it is not anti-constructionist because it does not deny that human beings are always a work in progress and that they are responsible for what they will make of themselves.

The "no nature" view arrogantly places humanity above nature—a pleasant enough thought, but a false one. We may be *Homo sapiens sapiens,* the doubly wise species, but we are still a subdivision of the primate order. The somewhat redundant second *sapiens* (modern humans are a subspecies of archaic *Homo sapiens*) perhaps marks (*pace* romantics) our ability to escape many of nature's restrictions and even to bend her to our will. Yes, humans are certainly unique, but we are not uniquely unique. Every animal species is unique in many ways vis-à-vis every other species. Indeed, every human being is in some way unique from every other member given the almost infinite number of possible permutations of genes and developmental trajectories that shape each person. However, as Aristotle may have argued, the variance in the *per accidens* components does not belie the central tendency of the *per se* whole.

To describe the nature of anything, we list its special features and nominate those that are unique, or quantitatively enhanced, that differentiate it from everything else that is not it. Because of training and temperament, the more radical wing of the political left remains the most insistent that there is no human nature, claiming that it is little more than the ensemble of social relations within a given mode of production. If this is so, then all we have to do to change human nature and achieve secular salvation is to change the mode of production. The left denies a universal human nature because it militates against its Platonic dream of human and social perfectibility, but it is this universal human nature that chafes to break free when megalomaniacs emerge who assume they can mold people to their ideological visions. The works of Karl Marx provided the philosophical justification for the visions of the likes of Stalin and Mao, and it is to Marx that leftist deniers of human nature most often turn for support.

On several occasions, however, Marx strongly implied a belief in human nature: "Man is directly a natural being . . . furnished with natural powers of life—he is an active natural being. These forces exist in him as tendencies and abilities—as instincts" (Marx 1978:115). For Marx the essential feature of any species is the activity that distinguishes it from every other species, and the primary distinguishing factor for humans relative to other animals is that they consciously create their environment rather than merely submitting to it. It is this creative activity, wrote Marx, that is the distinguishing feature of human *species being*, "man's spiritual essence, his human essence" (in Sayers 2005:611). This Aristotle-like human essence is the nature from which humans are alienated by capitalism, according to Marx. If Marx was right about this distinguishing feature of human nature, the component design features that allow us to create our environments must also distinguish us from other animals. These features include, but are not limited to, language, large brain size relative to body size, intelligence, rationality, self-consciousness, foresight, continuous sexual receptiveness, spirituality, a theory of mind (the ability to infer what others probably know, want, and will do), and moral sensibility. The sum of these and other feature may be rightly called human nature, and the relevant information for the design and expression of these features is contained in the species genome. Any geneticist given a sufficient number of alleles from the genomes of a variety of species will be able to differentiate humans from non-humans every time. This should suffice to convince anyone that humans are a "kind" marked off by the unique configuration of their DNA from all other animals that do not share that configuration—but obviously it does not suffice.

In *Capital*, Marx wrote, "To know what is useful for a dog, one must study dog nature. . . . Applying this to man, [we] must first deal with human nature in *general,* and then with human nature as *modified* in each historic epoch" (Marx 1967:609; emphasis added). Marx is boldly stating that there is a universal human nature as well as one modified by a particular mode of production, and any social theory must be predicated on an understanding of that nature. As I read Marx, he was positing three human natures: the biological nature we share with all humans everywhere; the social nature we share with those who share our culture, and the nature we make for ourselves as reflexive creatures of self-directed action. While I concur with this position, I emphasize that a universal human nature is the foundation, scaffolding, and bricks used in the construction of our social and personal natures.

Melford Spiro, a self-confessed former cultural determinist, realized this after studying vastly different cultures in Micronesia, Burma, and Israel. His struggles with the data led him to come "to the conclusion that I could not make sense of my findings so long as I continued to operate within the postulates of strong cultural determinism and relativism. . . . I could see no way of accounting for [my findings] short of postulating a pancultural human nature" (1999:8). The existence of a pancultural human nature is supported by a number of studies that have strongly confirmed the universality of the same personality constructs. Across at least 50 cultures with vastly different social, cultural, economic, religious, and political systems, many hundreds of thousands of subjects show the same features of personality; i.e., the same recurring regularities that correlate with the same behavioral outcomes (Costa, Terracciano, & McCrae 2001; McCrae & Terracciano 2005; Schmitt et al. 2008).

Human nature, like human will, is developed, expressed, and constrained contingently, and therefore we should not take human nature as we observe it instantiated in a particular culture as exhausting all that human nature is. It is beyond doubt that hunter/gatherers, agrarians, and industrial workers living in vastly different times and in greatly different cultures express their natures differently, but these expressions are variations on a common theme running through time and place. Sometimes it is useful to understand what something is by understanding what it is not. No humans have ever lived in a culture such as the one presented below by David Buss (2001:973), which starkly illustrates what human nature just cannot be:

> They live in peace and harmony and don't get in into social conflicts; sex roles are reversed, with women being masculine and aggressive, men being feminine; husbands don't care if their wives have sex with other men, and women don't care if their husbands give the bulk of the meat from the hunt to their lovers; they lack envy, jealousy, and avarice; men find older women who are grandmothers to be more sexually attractive than young women; they lack status hierarchies and are perfectly egalitarian; and they channel acts of altruism as much toward other people's children as their own.

If an anthropologist staggered out of the jungle with such a description, it would really make news. There are even some constructionists in anthropology who would believe and celebrate such a story precisely because it made no sense and would be incommensurate with their own culture. Spiro (1999:11) quotes a noted anthropologist who turned Horace's famous apothegm upside down to write, "My own group aside, everything human is alien to me." This anthropologist was declaring that Western anthropologists are incapable of understanding other cultures because there is no "psychic unity of mankind" to provide a basis for such understanding.

If this is the case, students wanting to gain penetrating insights into human nature should avoid cultural anthropology and head for the English department. Grosvenor (2002:434) tells us that the timeless lessons of literature provide ample evidence for a universal human nature, writing that "It is the existence of perennial traits that enables us to understand, for example, the motivations of characters in the plays of Shakespeare or Sophocles, even though they were written in times radically different from our own." Grosvenor is saying that if there is no universal human nature underlying cultural variation, then the stories from ancient and distant cultures would mystify us, but they do not. How is it that we can understand Antigone's struggles against King Creon to secure a decent burial for her brother; how is it that Plato's allegories resonate so strongly with us; why do we appreciate Odysseus's struggle for self-identity and Penelope's patient and faithful love; and how is it that across the ages military leaders and businessmen the world over have found much to learn in Sun-Tzu's *Art of War*, written in 6th-century BCE China? Perhaps the greatest of all British philosophers, David Hume, wrote long ago that if we would like to "know the sentiments, inclinations and course of life of the Greeks and Roman," we would be advised to "study well the actions of the French and English" (in Trigg 1999:83). Indeed, If human beings in all cultures at all times did not have the same hopes, aspirations, character traits, emotions, feelings, goals, needs, moral strengths and weakness—if culture was a more or less arbitrary selection from a grab-bag of possibilities—then different cultures would indeed have no common factors by which we could judge them. If the goal is again to be good non-judgmental liberals, let us ask how we can denounce the inhuman if we do not know what the human is?

Natural Selection and Human Nature

All humans share a common nature by virtue of a common evolutionary history and a common genome. The human genome is the chemical archive of millions of years of evolutionary wisdom accumulated by natural selection. Any functional genes that are currently part of our genome are there because they provided some sort of advantage to our ancestors in the pursuit of the shared goals of all life forms: survival and reproduction. Evolutionary approaches to behavior utilize the modern synthesis of natural selection and genetics to test hypotheses regarding the functional advantages conferred by these genes, or rather by the phenotypic traits that they underlie. Evolutionists are interested in distal "why" explanations rather than proximate "how" explanations. A proximate explanation for sex differences in dominance seeking and aggression, for instance, may be that the sexes have different levels of testosterone, whereas an ultimate evolutionary explanation would involve exploring the adaptive rationale for why sex differences in testosterone exist in the first place.

Evolution by natural selection is a trial-and-error process that changes a population's gene pool over time by the selective retention and elimination of genes as they become adaptive or maladaptive in their environments. The nature of any living thing is thus the sum of its design features that arose and promoted their increased frequency through an extended period of natural selection because they functioned to increase survival and reproductive success. It is important to realize that natural selection does not *induce* variation; it is a process that *reacts* to it by preserving favorable variants. Only the recombinations that occur during the division of sex cells during meiosis can result in new allele combinations, and only mutations can produce new alleles for natural selection to work with.[1]

Evolutionary accounts do not ignore culture, as often claimed; they simply remind us that "psychology underlies culture and society, and biological evolution underlies psychology" (Barkow 1992:635). Ultimate-level explanations complement, not compete with, proximate-level explanations because nature (genes) and nurture (cultural learning) constitute a fully integrated reciprocal feedback system. Genes and culture are both information transmission devices—the former laying the foundation (the capacity) for the latter, and the latter then influencing the former (for instance, what genetic variants are useful in this culture at this time?). If a novel trait emerges that happens to be useful and desirable in a given culture, those displaying the trait will be advantaged in terms of securing resources and mates, and thus the alleles underlying the trait will be preserved and proliferate in the population gene pool. For instance, Herbert Gintis (2003) has shown how genes underlying altruism can become a fixed part of the genome because it is culturally valued and thus fitness enhancing for those who demonstrate it.

Sophisticated new gene technology has revealed that the rate of genomic change has been about100 times greater over the last 40,000 years than it was during the five-million-year long Pleistocene, due largely to the greater challenges posed by living in ever-larger social groups: "[T]he rapid cultural evolution during the Late Pleistocene created vastly more opportunities for further genetic changes, not fewer, as new avenues emerged for communication, social interaction, and creativity" (Hawks et al. 2007:20757). Our most human characteristics evolved during the Pleistocene epoch, but we do not operate with brains forged exclusively during that epoch. A number of studies of hominid crania dating as far back as 1.9 million years show more robust increases in cranial capacity in areas with greater population density and in areas in which food procurement was most problematic—namely colder and most northerly areas of the globe (Ash & Galluop 2007; Kanazawa 2008). One study of 175 crania found that latitude was strongly related to cranial capacity (r = .61), but population density was more strongly related (r = .79). These authors (Bailey & Geary 2009:77) concluded that the burden of evolutionary selection has moved from "climactic and ecological to social." Among more modern humans, new genetic variations affecting the brain's structure and function have been discovered as it continues to evolve in response to new ecological and social conditions (Evans et al. 2005; Mekel-Bobrove et al. 2005).

Human nature is thus not a timeless and immutable Aristotelian *per se* essence when viewed from an evolutionary time scale. Because species undergo the process of evolution, we can rule out any feature or conjunction of features as *timelessly* necessary and sufficient for membership. All modern humans were once Africans, who were once Australopithecines who once shared a common lineage with modern great apes, who, even further back in the mists of time, were creatures who reproduced asexually—and so on all the way back to the Big Bang. But this philosophical rewind to the fact that we ultimately are all, in Carl Sagan's famous line, "made of star stuff" (1980:233), does not rule out the use of the current features to distinguish human from non-human. After all, we've been the way we are for a very long time. As Lisa Gannett (2010:367) asserts, we can rule out the strong essentialist *"really* real" concept of human nature of the philosopher and embrace the "merely real" of the biologist.

Sexual Selection and Male-Female Natures

Natural selection forges a sex-neutral human nature, and a second mechanism called sexual selection forges separate natures for males and females. We may thus say that there are two human natures— male and female (Davies & Shackleford 2008). The features shared by the sexes (physiology, behavior, traits, characteristics, motives, and desires) dwarf the features that they do not share, or share at different average levels. The differences that are most salient to the core of one's identity as male or female, however, are large (reviewed in Hines 2004). Darwinian feminists stress that to understand these differences we have to begin, as the mature Margaret Mead noted long ago, with "sex differentiated reproductive strategies" (1949:160).

Charles Darwin's theory of sexual selection was added to his theory of natural selection when he noted that while natural selection accounted for differences *between* species, it did not account for the often profound male/female differences *within* species. We should be able to explain most of the characteristics of any organism as functional adaptations—as the result of genes that have filtered across the generations because of the fitness benefits they confer. Yet there are many male traits that natural selection could not explain. As evolutionary feminist Griet Vandermassen (2004:11) explains, "Darwin posited sexual selection as a way to account for many conspicuous physical and behavioral traits in males. These traits are so energy demanding and so likely to make the animal vulnerable to predators, that natural selection would have normally selected them away in an early evolutionary stage." Natural selection cannot explain the bright colors and elaborate morphology of peacocks' tails, for example, in terms of some survival advantage because there is none; so why are they there and why do they survive? The simple answer is that they survive because peahens like them.

Sexual selection involves competition for mates and favors traits that lead to reproductive success regardless of whether they have survival advantages. The peacock's bright plumage attracts females by indicating "good genes," but they are costly and invite easy predation. Biological fitness is a quantitative measure of reproductive success, not survival. Traits become amplified because they increase fitness by increasing the probability that their carriers are more likely than others to mate. Survival per se means nothing in evolutionary terms unless the organism passes on its genes. Sexual selection, like natural selection, causes changes in the relative frequency of alleles in populations in response to environmental challenges—but in response to sex-specific mating challenges rather than general sex neutral ecological challenges (Qvarnstrom, Brommer, & Gustafsson 2006). Males and females inhabit the same ecological niches in which natural selection operates in sex-neutral ways, but they inhabit different mating environments that ultimately produce different sex-based natures.

The idea of different sex-based natures begins with the principle of anisogamy, which refers to the different biological values of the male sperm and the female egg. Sperm cells lack nutrients and are little more than packets of chromosomes with tails made by the billions every day. Egg cells are about 85,000 times bigger than sperm cells, with typically only one released each month from puberty to menopause (about 500 in total), and are rich in nutrients (Bateman & Bennett 2006). Given the rarity of egg cells, a female's unconscious imperative is to choose wisely which male she will allow to fertilize them. With cheap and plentiful sperm cells, males can be profligate in their expenditure (Campbell 2009).

There are two paths by which sexual selection proceeds: Intrasexual and epigamic selection. Intrasexual selection is primarily male competition for access to females, and epigamic selection is a process of females choosing with whom they will mate. The more intrasexual selection operates on a species, the more sexual dimorphism there will be. In species where intrasexual selection is paramount, there are large differences between males and females in size, strength, and aggression, and in species where epigamic selection predominates, males are more striking in their appearance than females (Andersson & Simmons 2006). Large degrees of sexual dimorphism reflect a polygynous mating history in which dominance is established by physical battles among males. The earliest hominid males (Australopithecines) were 50% to 100% larger than females (Geary 2000). The fairly low degree of sexual dimorphism for body size in *Homo sapiens* (modern men are only about 15% larger than women, on average) indicates an evolutionary shift from violent male competition for mates to a more monogamous mating system and an increase in paternal investment (Plavcan & van Schaik 1997).

Selection for Biparental Care and Its Consequences

The evolution of biparental care and monogamous mating patterns among humans put the skids to runaway sexual selection, which is why we see neither enormous size nor costly decorations in human males. Biparental care is found only in about 10% of mammalian species and is found in species in which offspring remain highly dependent for a long time, for which food procurement is somewhat problematic, and in which rates of predation are neither too high nor too low (Manica, & Johnstone,2004). Pair bonding is selected for only when the help of a male positively influences the probability of offspring survival by procuring food for gestating and lactating mothers and by defending mother and child against predation. In precocial species with ready access to food, and with predation rates so high or so low that male parental investment is unlikely to have any positive effect on offspring survival, pair bonding is not necessary, and no evolutionary pressure is exerted for its selection (Quinlan & Quinlan 2007). Given the long period of human infant dependency, it makes sense for ancestral females to choose mates inclined to invest resources in offspring. As Campbell (2004:17) put it: "Monogamy may have been the result of male-female coevolution of reproductive strategies, initiated by female preference for investing males." The logic of epigamic sexual selection points to the conclusion that the evolution of many male traits and behavior was driven by females; human males are the way they are because our ancestral foremothers liked them that way.

We have noted that humans have been on a steep evolutionary trajectory toward greater intelligence from Australopithecines to modern *Homo sapiens*. Greater intelligence requires a bigger brain to lug it around. Brains are particularly voracious in their appetites for energy, thus selection for increased brain size would result only from extreme pressures, and the human brain is much larger than should be reasonably predicted for a species of our body size (Dunbar & Shultz 2007). The increased cranial size to store bigger brain mass placed tremendous reproductive burdens on females. The human birth canal cannot accommodate birthing infants whose brains are 60% of its adult weight, as in newborn macaques, or even 45%, as in newborn chimpanzees (Hublin & Coqueugnoit 2006). The pelvis of Australopithecine females was probably shaped to accommodate upright posture and bipedalism (which has the effect of narrowing the birth canal) more than to accommodate increased infant skull size, thus precipitating a conflict between the obstetric and postural requirements of ancestral females (van As, Fieggen & Tobias 2007). Evolutionary conflicts such as this are not uncommon; natural selection works on trajectories already in motion, and it cannot anticipate future needs.

Evolution partially solved the obstetrics/posture conflict (human females have more difficulty giving birth than other species because of this problem) by having infants born at ever earlier stages of development as cerebral mass increased. Human infants experience 25% brain growth inside the womb (*uterogestation*) and 75% outside (*exterogestation*) the womb (Perry 2002). The high degree of developmental retardation of the human brain assures a greater role for the extra-uterine environment in its development than is true of any other species.

A species burdened with extremely altricial young such as ours will experience strong selection pressures for neurohormonal mechanisms designed to assure the young would be nurtured for as long as necessary. The long period of dependency required selection for strong bonds (attachment) between mother and infant, and the extra caregiving demand on females produced selective pressure for male/female bonding. Because mother/infant and male/female bonds involve an active concern for the well-being of another and share crucial evolutionary goals, both forms of bonding share a common neurobiology (Esch & Stefano 2005). Attachment mediating neurohormones activate regions in the brain's reward system specific to either maternal or romantic love, but there are also large overlapping regions that are activated by both types of love (Bartels & Zeki 2004). Male/female bonding probably originated with females choosing to mate with males who showed a penchant for sharing and caring rather than the more dominant and aggressive males who loved and left. Males and females who bonded to jointly provide parental investment increased the probability of their offspring surviving to reproductive age, thus improving their own reproductive success (Campbell 2004).

Because sexual selection provides the understanding of the evolutionary basis of sex differences, there is bound to be some opposition to it from certain strands of feminism. Joan Roughgarden (2009), for instance, abhors sexual selection theory because she has a passionate devotion to non-traditional sex roles and non-traditional gender identities. Incredibly, she declares that our ancestral females usurped parental care from "naturally monogamous" males and drove them into promiscuous competition for mating opportunities "as a tactic of last resort" (2009:190). Straying cads may now cite Roughgarden in defense of their restless search for more wombs to carry their seed: "It's all the fault of out foremothers—I'd much rather be home nurturing the young 'uns than dallying with yon wench."

Her caricature of one of the best established theories in all biology leads her to resurrect a dichotomy many believed to be long dead: nature versus nurture. She proposes something called "social selection theory" in opposition to sexual selection theory, thus splitting asunder the nature-nurture whole we have been discussing. Not to be outdone by others claiming to have discovered the vile sexist underbody of science, or by those who love the rape metaphor, she excoriates sexual selection theory as "rape in scientific guise, a narrative of males victimizing females" (2009:103). She writes this despite sexual selection theory's assertion that the whole process is driven by either intrasexual competition in which males "victimize" other males in their competition for mating opportunities, or by epigamic selection in which females choose the male with whom they will mate.

The error in opposing the social to the biological in evolutionary thinking, as Roughgarden does, is exemplified by studies showing how cultural practices can lead to morphological changes. Studies of primate brains have shown that group life has produced striking differences between the sexes in brain mechanisms related to carrying out the different demands placed on males and females in evolutionary environments (Dunbar 2007). Because of the competitive demands of sexual selection for males, we should observe greater development of subcortical (limbic) brain structures involved in sensory-motor skills and aggression. Conversely, female fitness depends more on acquiring male resources and navigating social networks, and thus we should expect greater development of neocortical areas, particularly of the frontal lobe structures (Lindenfors 2005; Lindenfors, Nunn, & Barton 2007). Patrik Lindenfors and his colleagues (2005; 2007) have shown this to be the case among 21 non-human primate species ranging from chimpanzees to rhesus monkeys. They found that the more affiliative sociality of females is related to greater neocortex volume and that the more competitive male sociality is more closely related to subcortical (limbic system) volume.

The Lindenfors team suggests that this should extend beyond the primate species studied so far, including our own. Indeed, a functional magnetic resonance imaging (fMRI) study showed a greater ratio of orbital frontal cortex volume to amygdala volume in human females relative to males (Gur et al. 2002). The orbital frontal cortex is part of the prefrontal cortex (PFC). This vital part of the human cortex has extensive connections with other cortical regions and with deeper structures in the limbic system. Because of its many connections with other brain structures, it is generally considered to play the major integrative as well as a major supervisory role in the brain, playing vital roles in forming moral judgments, mediating affect, and for social cognition (Romain & Reynolds 2005). The amygdala's primary function is the storage of memories associated with the full range of emotions, particularly fear.

Conclusion

Like everything that exists in the phenomenal world, humans have a nature. To be sure, it is a malleable nature, but malleability is part of the definition of what human nature is. We are designed to incorporate information from diverse sources into our neural architecture in somatic time and into our genomes in evolutionary times, and thus we change. We change and adapt to environmental challenges, and "Human nature may be defined as our collection of adaptations"

(Kennair 2002:27). Our human nature is Gannett's (2010) "merely real" of biology and not the immutable and timeless "really real" of philosophical essentialism. It seems to be a ploy of those seeking to deny the existence of something or other to define it according to impossibly strict criteria and then to say that it either exists in the way they have defined it or it does not exist at all. The deniers of human nature assert that we are entitled to speak of it (or anything else with a putative nature) only if properties used to define it are timeless and immutable. In this sense they are asking for a human nature as pure as a Platonic Form and as provable as a mathematical axiom. If supporters of human nature cannot do this, then human nature does not exist. Thus is the position of those who make their living either affirming or denying the obvious.

The Darwinian theories of natural and sexual selection are the meta-theories underlying our current knowledge of what human nature is and how it got to be that way. This observation should be patently obvious to everyone, given that evolution is the one and only organizing principle for explaining why all animals are the way they are and do what they do. Thus, it is no surprise that Griet Vandermassen (2004) urges the adoption of sexual selection and parental investment theories to move feminism to a higher intellectual plain. These theories, she avers, allow us to predict gender-related roles and behavior purely on a priori theoretical grounds, which no other theory of behavior can do: "Theories of socialization, for instance, can only predict how gender roles will affect people if one already knows what these roles are. They cannot explain why the same gender differences are reliably found all over the world" (2004:20). Vandermassen is concerned with feminism, but a Darwinian view of all things human has the potential to move the whole enterprise of behavioral science forward.

Having examined the nature and tools of science and social constructionism, and the numerous disagreements between them, we can now turn our attention to the main substantive issues in the contemporary science wars: gender and race. In a nutshell, the arguments about gender revolve around the extent to which gender differences map to biological sex differences, and for race, the issue is its very existence. One would think that such issues could be easily addressed and resolved with the technology available to science today, but so many fears and concerns stand in the way that the wars surrounding these issues may last a long time.

Endnote

1.   There is a pervasive misconception about Darwinian natural selection that tends to alienate many humanists. This misconception arises from the use of the phrase "struggle for existence." This phrase has been interpreted in light of Alfred Lord Tennyson's "Nature, red in tooth and claw," indicating that evolution is all about aggressive intraspecies competition for scarce resources. But this is not what Darwinism is about. Darwin wrote about cooperation as the major strategy for survival and reproductive success, and mentioned cooperation three times more than conflict (Levin 2006). Darwin himself felt it necessary to explain what he meant by the struggle for existence: "I should premise that I use the term 'Struggle for Existence' in a large and metaphorical sense, including dependence of one being on another, and including (which is more important) not only the life of the individual, but success in leaving progeny" (in Lieberman 1984:4).

Chapter 6

Social Constructionism and Gender

Gender Construction

The concept of gender is the cornerstone of feminism. The term gender has traditionally been used to refer to the masculine and feminine grammatical categories of language, but sometime in the 1960s it came to be used to refer to categories of human males and females (Nicholson 1994). The use of gender as a synonym for sex in everyday life is useful to avoid connotations of copulation. It is also useful for gender or equity feminists, who can point to the arbitrary gender categorization of nouns in different languages (e.g., table is masculine in French and feminine in Spanish) and then slither into the argument that gender as it applies to people is also arbitrary.

For social scientists the terms sex and gender refer to different but overlapping concepts. Sex refers to a person's biological status as a male or female, while gender refers to social or cultural categories about how femininity and masculinity are culturally molded and expressed. Sex, with some rare exceptions, is universal in its sameness, but gender is a fluid and dynamic social construction built upon the superstructure of sex, although the more extreme social constructionist may not accept that gender is based on anything material. Radical gender constructionists believe that gender differences are real only to the extent that the consensus in society believes them to be. Nor do these folks believe that there is any logic by which we could predict patterns of behavior from a person's chromosomal sex. Unlike sex differences that exist independently of our ideas about them, the existence of gender differences depends to some extent on a culture's shared discourse, but they are not independent of biological sex. Gender-appropriate behaviors, activities, and mannerisms are in certain ways like clothing or musical fashions that flit in and out of favor arbitrarily across different cultures, subcultures, and historical periods, but they do not stray too far from the template of biological sex in any culture or historical period.

Some constructionists even claim that sex differences are social constructions. Judith Lorber (1994:46), for instance, claimed that "a purely biological substrate [of gender] cannot be isolated because human physiology is socially constructed and gendered." Lopreato and Crippen (1999:143) responded to this gem of postmodern wisdom by stating, "One wonders whether members of the medical profession are aware of this extraordinary discovery." Others have maintained that such things as pregnancy sickness and menstruation pains are socially constructed and that because females are fed less and have fewer exercise opportunities, they are smaller and weaker (Minkola 2008). Minkola (2008:13) goes on to opine that "if males and females were allowed the same exercise opportunities and equal encouragement to exercise, it is thought that bodily dimorphism would diminish." Thus, gender is not created from anything material (e.g., DNA, hormones, brain cell tissue), but entirely from insubstantial social stuff such as cultural attitudes, values, and role expectations. It is from such material that Bruce Charlton (2010) constructed his "clever sillies" concept.

From whence do such bizarre opinions come? I have previously commented on the seductive power of constructionism as lying in what it permits us to claim; biologist and feminist Helena Cronin (2003:59–60) adds similar thoughts on the matter:

> It's as if people believe that if you don't like what you think are the ideological implications of the science then you're free to reject the science – and to cobble together your own version of it instead. Science doesn't have ideological implications; it simply tells you how the world is – not how it ought to be. So, if a justification or a moral judgment or any such "ought" statement pops up as a conclusion from purely scientific premises, then obviously the thing to do is to challenge the logic of the argument, not to reject the premises. But, unfortunately, this isn't often spelled out. And so, again and again, people end up rejecting the science rather than the fallacy.

An almost inevitable consequence of the embrace of strong social constructivism and the repudiation of science are positions such as those of Dorie Klein (1995:50), who asserts that "It is not the existence of the two genders that generates sexism but the other way around; in other words, women and men are not just made, but made up. . . . That we divide humans into two genders is a social artifact." How we could divide humans (or any other species) into anything but the binary categories of male and female is not revealed. Perhaps she is referring to how a person with one set of genitals or the other subjectively self-identifies and behaves—how they express their masculinity or femininity. If this is what she means, then I can agree with her that this felt sense of gendered identity is culturally conditioned, but it is also biologically conditioned. Similarly, Barrie Thorne (1993:2) argues, "Parents dress infant girls in pink and boys in blue, give them gender differentiated names and toys, and expect them to act differently. . . . In short, if boys and girls are different, they are not born, but *made* that way."

Feminists such as Klein and Thorne claim that sex-differentiated socialization practices around the world are entirely arbitrary historical accidents because they have no biological foundations that could direct them in any predictable ways. Barbara Ehrenreich and Janet McIntosh (1997:12) call this "gender from nothing" position the "new secular creationism" that threatens the credibility of feminism. It is vital that social scientists who still hang on by their fingertips to the idea that nature and nurture are mutually exclusive categories let go and grasp the firm overhang of science. Nature needs nurture if it is not to become a chaotic tangle of rank weeds, and nurture needs nature because without it nurture has no place to go.

Gender Socialization and Gender Differences

Let us all agree that gender socialization means learning gender behavior, attitudes, and roles considered to be appropriate in one's culture, and is imparted by the family and reinforced by friends, school, work, and the mass media. Parental expectations of boys and girls lead them to buy gender-specific toys and assign them gender-based tasks. Thus, children learn to behave in ways dictated by societal beliefs, values, attitudes, and examples about how the sexes ought to present themselves. Children are exposed to norms that define masculine and feminine from an early age. Boys are told not to cry, not to be afraid, and to be assertive and strong. Girls are permitted to cry and are encouraged to be ladylike.

The school curriculum reinforces cultural ideas of how girls and boys should act in the course material (Dick and Jane) and activities that separate boys and girls. Teachers are said to reinforce gender roles by encouraging girls to excel in the verbal world and boys to excel at math and science, and they tend to praise boys for substantive content and girls for neatness. Children absorb these lessons and develop an "us and them" attitude, dividing themselves along gender lines in the lunchroom and playground. Boys who play with girls are called sissies, and girls who play with boys are called tomboys. Peers are thus powerful socializing agents conditioning each other to adopt the correct gender attitudes and behaviors.

Children become aware of their sex by about age 2½, and by 3½ they become aware that the world divides people according to male and female (Lippa 2002). When this awareness strikes them, they embark on a subconscious program of self-socialization ("I'm a boy/ girl, and this is how I should and will act"). Children soon develop a gender-related self-concept and seek understanding of what gender behavior is appropriate for them. This is a part of their growing general awareness of how their world is organized and how they and others fit into it, and is known as a gender schema. The development of a gender schema is a process by which incoming information is labeled and boxed as male- or female-typed, which then becomes the basis for enhanced attention and concern about gender-appropriate behavior.

If gender schema building is thought to proceed without a biological foundation, the construct will not hold up. Cracks began to appear in its edifice when a series of longitudinal studies showed that young children preferred to interact with members of their own sex and preferred sex-congruent toys before they were able to label behavior or toy preferences as male- or female-appropriate, or even sort pictures of girls and boys into correct piles (Campbell, Shirley, & Candy 2004; Trautner 1992). As Anne Campbell (2006:79) put it, "Children seem to need neither the ability to discriminate the sexes nor an understanding of gender stereotypic behavior to show sex differences." Efforts to get children to play with opposite-sex toys or to play in less segregated groups are strongly resisted: "Even when adults try to encourage cross-sex play groups, children resist and quickly return to same-sex partners when adult supervision is reduced" (McIntyre & Edwards 2009:87). The evidence seems to tell us that what some dismiss as gender stereotypes are in fact reasonably accurate assessments of gender differences: Differences lead to stereotypes; stereotypes do not lead to differences.

In addition to primary socialization, social expectations are imparted by society through the social roles it deems sex appropriate. Social roles underpin what expectations society has for the sexes, and if these expectations are different, the argument is that the behavior and attitudes of the sexes will be different. If male roles demand independent, assertive, and competent behavior, males will tend to develop traits congruent with those demands. Conversely, if female roles demand communal, expressive, and nurturing behavior, that is what they will become. Social role theory avers that it is the role that molds the behavior, traits, and inclinations of men and women and not the other way around. A corollary of this is that as strict sex-based roles diminish, the sexes should converge in their psychological attributes. A specific prediction is made by Eagly and her colleagues (2004:289) when they claim that the "demise of many sex differences with increased gender equality is a prediction of social role theory."

Sex role theory would therefore predict that gender differences should be strongest in more traditional and patriarchal cultures where sex roles are most distinct. Contrary to this expectation, Costa, Terracciano, and McCrae's (2001) study of gender differences in personality across 26 cultures (n = 23,031) showed that gender differences were most pronounced in modern egalitarian cultures in which traditional sex roles are minimized. Another study (McCrae & Tarracciano 2005) using different measures and 50 cultures (n = 11,985) found exactly the same thing. A study by Schmitt and colleagues (2008) of 55 cultures (n = 17,637) of gender differences in personality traits found the exact same pattern; the biggest gender differences were found in cultures where sex role differences are minimized. Finally, Merten's (2005) study of emotional reactions among 42,638 participants in a variety of cultures once again showed that gender differences increase as gender equality increases.

The fact that all studies of this kind find the direct opposite of what sex role theory predicts is a shock to constructionists, but biosocial theorists are not surprised. The studies fit with behavior genetic studies of all kinds of traits, behaviors, and attitudes showing that as the environment becomes more equal for the traits in question, the more innate factors contribute to variance in the trait. There are only two sources of trait variance—genetic and environmental—so the more one source is equalized, the more the influence of the other stands out. This is a simple mathematical truism. Another way of looking at it is to say that in less constraining environments of modern egalitarian societies, individuals are freer to be themselves—that is, to construct their environments in ways consistent with their innate proclivities. Increased sexual dimorphism in personality in developed societies is simply a function of the natural tendency of males and females to develop different traits and personalities. Schmitt and his colleagues (2008) maintain that traditional agrarian cultures with their typically extreme levels of resource and gender inequalities may represent the largest departure from the egalitarian hunter-gatherer cultures that characterized our species for more than 99.9% of its history. Western post-agrarian cultures are closer to our egalitarian hunter-gatherer psychology, and thus we are freer to develop as we are naturally inclined (Adkins & Guo 2008).

I am not gainsaying the power of socialization. Without socialization we would all be barbarians: isolated feral creatures bereft of identity, morality, and direction, moved only by vague feelings and emotions. Socialization is the dialectic process by which the world and its meanings are filtered to us and can turn us into chivalrous knights or uncouth peasants, atheists or fanatical jihadists, tinkers, tailors, soldiers, sailors, beggars, or thieves. It cannot, however, change boys into girls or girls into boys. Nothing in biosocial science counsels ignoring the power of socialization. It only counsels against granting it exclusive rights to explaining human behavior, particularly behavior that impacts the overwhelming concern of all living things: survival and reproductive success. Except for occasional small changes of veneer, these concerns are simply too important to leave to the vagaries of learned cultural practices.

If gender identity is constructed solely by expectations and training, we would not find individuals socially constructed as gays and lesbians rejecting their lifelong heterosexual lessons in favor of their own privately constructed identities. By privately constructed, I mean that homosexuals are conforming to the way their brains, genes, and hormones bias them, despite social pressures to the contrary. These pressures have included religious and family exhortations, psychotherapy, hysterectomy, lobotomy, imprisonment, and even the threat of death, to coerce them into heterosexuality (Walsh & Hemmens 2011). Heterosexual males and females also construct their identities in conformity with their biology as well as in conformity to their cultures, and we can no more change that than we can change a homosexual's identity. I am aware that sexual identity and gender identity are two different (but overlapping) things. I use this example only to provide an instance showing the inability of a variety of cruel social pressures to change what is apparently innate.

The Influence of Margaret Mead

If gender differences were arbitrary social constructs decoupled from biology, we should observe three different gender outcomes randomly distributed across the historical and anthropological record. Under random (arbitrary) conditions, simple probability tells us that about one-third of cultures would have decided to socialize the sexes in traditional ways (i.e., males to masculinity, females to femininity). One-third of the cultures would have decided to reverse this process and socialize males in feminine ways and females in masculine ways. In this latter case, females would rule the roost, beat their men, kill each other in their matriarchies at rates men do in their patriarchies, take every opportunity to bed any male that came along, and march off to war while their men stayed home knitting socks and planting victory gardens. Finally, in about another one-third of the cultures, socialization would be androgynous, and neither sex would have cause to complain about the other.

The youthful Margaret Mead's book *Sex and Temperament in Three Primitive Societies* (1935) purported to show precisely such an arbitrary socialization model. Mead studied the Arapesh, Mundugumor, and Tchambuli peoples of New Guinea and reported that the Arapesh were a gentle, cooperative people who extolled a feminine temperament as the ideal for both sexes (androgynous feminine). The warlike Mundugumor people took the masculine temperament to be suitable for both sexes (androgynous masculine). But it was the Tchambuli who really excited those who believe in such things, because they supposedly turned gender upside down by favoring a feminine temperament for males and a masculine temperament for females. Mead (1935:279) reported that her "discovery" among the Tchambuli was "a genuine reversal of the sex attitudes of our culture, with the woman the dominant, impersonal, managing partner, the man the less responsible and emotionally dependent person." She further wrote that, "We are forced to conclude that human nature is almost unbelievably malleable, responding accurately and contrastingly to cultural conditions" (1935:289).

Mead was a glib player who rode this scruffy research nag a long way along a foggy ideological trail. She rode the poor thing to the social science glue factory, where others stuck together reams of arid prose that has delighted constructionists for more than three-quarters of a century. Her work took on the aura of a sacred text among feminists and social constructionists and was cited widely and approvingly in anthropology and sociology textbooks well into the 1990s (Roscoe 2003). It is less frequently found today, however, because a number of curmudgeons demanded to actually inspect the basis of her claims, and when they did, the glue cracked and the pages crumbled.

Generations of gullible students have been exposed to this research despite the fact that criticism of it since the 1930s has told us that sex-based temperament, behavior, and roles in these cultures were not too different from what they are in cultures around the world. Fortune's (1939) study of the Arapesh pointed out that they did not expect the two sexes to have the same temperament, that boys could only be initiated into manhood after they had committed homicide, and that warfare was a well-developed art among these "gentle, feminine" people: "Violence and war were very much a part of their established tradition" (Roscoe 2003:589).

Mundugumor males were indeed warlike and violent, as Mead described, but their women expressed their supposedly "masculine" aggression mainly by striving to please their men in ways that upstaged their co-wives, and their menfolk thoroughly dominated them. Deborah Gewertz's (Gewertz 1981; Gewertz & Errington 1991) fieldwork among the Tchambuli showed it to be a thoroughly male-dominant society where aggressive behavior on the part of the supposedly in-charge women earned them a beating from their "feminine" husbands. It was true that Tchambuli males were vain and took an almost narcissistic pride in their ritualized appearance, particularly their war paint, which Mead mistakenly took for feminine "makeup" (Roscoe 2003). Countless macho males around the world wear their versions of war paint (tattoos, gang colors, badges, uniforms) and preen and strut their stuff, but no one calls them feminine to their faces without some considerable risk.

This is not to say that gender is not variable. Different cultures can and do mold masculine and feminine characteristics in diverse ways, and there is considerable overlap between the sexes/genders on many traits. However, trait variability among individuals, male or female, is attributable mainly to genetic factors (Craig, Harper, & Loat 2004). That is, gender socialization interacts with individual temperaments to produce the ways individuals "do gender." Genetic studies find only miniscule effects of shared environment (which includes everything siblings shared as children, including parental socialization practices) on gender roles, although they do find considerable non-shared environmental effects (McIntyre & Edward 2009).

Ecological Explanations

Social scientists would tell us that culture explains the different behaviors of these peoples and leave it at that, thus begging the question of what lies behind these cultures that makes them different. Many modern anthropologists realize that cultural practices are underlain by the nature of their physical environment (ecology). Ecology necessitates certain behaviors that are then reinforced by cultural practices. David Lipset (2003:699) notes (as did Mead) that the Arapesh lived in harsh mountain conditions characterized by chronic food shortages, while the Mundugumor occupied "fertile grounds, divided by small channels that were full of fish." The different ecological niches in which the Arapesh and Mundugumor lived may be sufficient to explain their different cultural temperaments. In the ecological tradition, Harpending and Draper (1988) contrasted the reproductive strategies of the !Kung bushmen and the Mundurucu. The !Kung inhabit the inhospitable Kalahari desert in South Africa, and the Mundurucu inhabit the resource-rich Amazon basin. Because conditions are harsh in the Kalahari, life is precarious, cooperative behavior is imperative, and feminine parenting effort is favored over masculine mating effort among the !Kung, as it is among the Arapesh. The Mundurucu's rich ecology frees males for fighting, for raiding other groups, and for engaging in fierce competition for females. Mating effort is thus favored over parenting effort among the Mundurucu and the Mundugumor for ecological reasons.

As long as social scientists view culture as an autonomous causal agent containing a more or less arbitrary grab bag of roles, values, and customs, we can never understand much about group differences in behavior. A coherent explanation of cultural differences requires an understanding of human nature and the fitness imperatives imposed on it. The peoples of all five cultures have similarly evolved adaptations, but they are constrained to execute those adaptations in different environments. Only an understanding of human nature can help us to appreciate the different psychologies underlying the social behavior of people in these five cultures in ways that would lead to predictions about the behaviors of other groups inhabiting similar ecological niches. Culture is important in explaining variation in human behavior, but it is not a realm ontologically distinct from biology.

The mature Margaret Mead came to acknowledge the biasing framework supporting her "temperaments" (she seemed to have had gender in mind before the term was invented with her use of the term temperament). In her later work *Male and Female* (1949) she wrote, "If any human society—large or small, simple or complex, based on the most rudimentary hunting and fishing, or on the whole elaborate interchange of manufactured products is to survive, it must have a pattern of social life that comes to terms with the differences between the sexes" (1949:173). She traced these differences to "sex differentiated reproductive strategies" (1949:160). As for her youthful claims about the Tchambuli, she later remarked, "All the claims so glibly made about societies ruled by women are nonsense. We have no reason to believe that they ever existed. . . . men everywhere have been in charge of running the show" (in Goldberg 1986:31). Such statements tend to raise hackles among constructionist feminists regardless from whose pen they come, but it marks the mature Mead as a scientist who followed the data wherever they led her, despite her earlier claims to the contrary.

Melford Spiro: The Reluctant Apostate

Melford Spiro's (1975, 1980) studies of Israel's kibbutzim were perhaps the most devastating blows to the gender-as-social-construct argument to come from the social sciences. The kibbutzim movement provides us with a natural experiment that could never be artificially duplicated by scientists. Begun in 1910 and heavily influenced by the Marxism of Russian immigrants, one of the purposes of the communal movement was to strip its members of all vestiges of bourgeois culture and to abolish sex-segregated social roles. Boys and girls were raised collectively, taught the same lessons, given equal responsibilities, and shared the same toys, games, living quarters, toilets, dressing rooms, and showers. This sex-neutral socialization was supposed to result in androgynous beings devoid of observable differences in nurturance, role preferences, empathy, aggression, or any other trait or behavior said to be sex linked.

In commencing his studies, Spiro thought that he was setting out to discover and document the changes in human nature brought about by the movement because he was a social constructionist about human nature. What he actually found forced on him what he describes as "a kind of Copernican revolution on my own thinking" (1980:106). "As a cultural determinist," he wrote, "my aim in studying personality development in 1951 was to observe the influence of culture on human nature or, more accurately, to discover how a new culture produces a new human nature. In 1975 I found (against my own intentions) that I was observing the influence of human nature on culture" (1980:106).

Spiro found a counterrevolutionary feminization of the *sabra* (kibbutzim born and reared) women. Despite decades of sex-neutral socialization and the exhortations of their ideologically committed foremothers, *sabra* women fought for formal marriage vows, greater contact with their children, and for separate toilets, showers, and living arrangements prior to marriage. In an earlier work, Spiro (1975) found that the activities and fantasy lives of young children varied significantly between the sexes despite strenuous efforts to eliminate them. Summarizing his data, Spiro wrote, "sexually appropriate role modeling is a function of precultural differences between the sexes" (1980:107). *Precultural*, of course, means an innate human nature.

In common with Margaret Mead, Spiro was eventually dragged by his data to embrace biosocial explanations for gender differences, as well as a universal human nature. In his presidential address to the Society for Psychological Anthropology (1999), Spiro detailed his intellectual journey from what he called "strong cultural determinism" and "strong cultural relativism" to a more nuanced bio-psycho-social view of human behavior. In his own words (1999:10),

> Having become increasingly disenchanted with, and bored by, the conceptual poverty of ethnographic particularism, and its mantra-like invocation of cultural determinism (now "cultural constructionism") to explain virtually everything—and hence nothing—the work of this group [the Society for Psychological Anthropology] opened my eyes to new and exciting explanatory vistas.

Conclusion

The gender-as-social construct argument has been examined and dismissed by looking at a variety of lines of evidence. We cannot continue to infer the power of gender-differentiated norms from the gender-differentiated behavior that these norms supposedly explain. Richard Udry (1994:563) calls such reasoning circular and states that "The reason for this tautology is that we, as social scientists, can't think of any other way to explain sex differences." Alice Rossi certainly thought of other ways in her 1984 presidential address to the American Sociological Association. In her address she warned her colleagues that if they continued to rely on disembodied phenomena to explain sex/gender differences they would become irrelevant in the scientific world:

> Gender differentiation is not simply a function of socialization, capitalist production, or patriarchy. It is grounded in a sex dimorphism that serves the fundamental purpose of reproducing the species. Hence, sociological units of analysis such as roles, groups,

networks, and classes divert attention from the fact that the subjects of our work are male and female animals with genes, glands, bones and flesh occupying an ecological niche of a particular kind in a tiny fragment of time. And human sexual dimorphism emerged from a long prehistory of mammalian and primate evolution. Theories that neglect these characteristics of sex and gender carry a high risk of eventual irrelevance against the mounting evidence of sexual dimorphism from the biological and neurosciences. (1984:1)

Alice Rossi was a former strict environmentalist and a card-carrying liberal with impeccable feminist credentials. As a founding member of the National Organization of Women, she was no tool of the patriarchy. She simply pointed out that gender differences arose from fundamentally different reproductive roles of males and females that have been fine-tuned by eons of evolutionary selection pressure. She was also stressing that sociologists must integrate the hard data from the more robust sciences into their work if they and their theories are to attain credibility within the broader scientific community.

Chapter 7

The Neurohormonal Basis of Gender

Sexing the Brain

The neuroscience explanation of gender differences rests on a foundation of differential neurological organization shaped by a complicated mélange of prenatal genetic and hormonal processes that reflect sex-specific evolutionary pressures. Meta-analyses have shown that most measured gender differences are small and relatively inconsequential, but the differences that are most salient to core gender identity—in other words, at the center of one's identity as male or female—are very large and have neurohormonal underpinnings (Hines 2004; 2011; Lippa 2002). The issue is thus not whether male and female brains are similar or different, because they are both. The interesting questions lie not in similarity but in differences, the ways in which they are different, and what these differences mean for gendered personalities and behaviors.

Doreen Kimura informs us that sexual selection pressures assure that males and females arrive in this world with "differently wired brains," and these brain differences "make it almost impossible to evaluate the effects of experience independent of physiological predisposition" (1992:119). Sarah Bennett and her colleagues (2005:273) concur, and assume that these brain differences lead to behavioral differences:

> Males and females vary on a number of perceptual and cognitive information-processing domains that are difficult to ascribe to sex-role socialization. . . . the human brain is either masculinized or feminized structurally and chemically before birth. Genetics and the biological environment in utero provide the foundation of gender differences in early brain morphology, physiology, chemistry, and nervous system development. It would be surprising if these differences did not contribute to gender differences in cognitive abilities, temperament, and ultimately, normal or antisocial behavior.

Jeanette Norden (2007:117) tells us that "Male and female brains can be distinguished on the basis of how particular structures are organized at gross, cellular, or even molecular levels," and fellow neuroscientists De Vries and Sodersten (2009:589) agree: "Thousands of studies have documented sex differences in the brain in practically any parameter imaginable." These differences begin with the genes encoded on the sex chromosomes specifying an XX female or an XY male: "These genes are differentially represented in the cells of males and females, and have been selected for sex-specific roles. The brain is a sexually dimorphic organ and is also shaped by sex-specific selection pressures" (Arnold 2004:1).

Sex differentiation begins with the Y chromosome, a puny little creature that probably evolved from a pair of autosomes in ancestral vertebrates many millions of years ago (Xu & Disteche 2006). The male-specific Y chromosome has only 27 genes coding for proteins on it, whereas the X, shared by both males and females, encodes for about 1,500 (Arnold et al. 2009). Further emphasizing Y's genetic poverty is the fact that most genes on the Y have homologous genes on the X chromosome, which shrinks the functional differences between XY and XX cells (Arnold et al. 2009). This leaves precious few genes that are male specific, and perhaps only one that really is, and that is the SRY ("sex determining region of the Y chromosome") gene.

In all mammalian species, maleness is induced from an intrinsically female form by processes initiated by the SRY gene at around the sixth week of gestation. All XY individuals would develop as females without the SRY gene, and XX individuals have all the material needed to make a male except this one gene. The major function of the SRY gene, and its downstream genetic cohorts, such as the autosomal SOX9 gene and the sex chromosomal DAX-1 gene, is to induce the development of the testes from the undifferentiated gonads rather than the ovaries that would otherwise develop in its absence. The SRY gene is a necessary but not sufficient step toward masculinization, and it is not all-powerful. There are rare cases when a double dose of the DAX-1 gene on a male's X chromosome, a single copy of which is part of the DNA team that build the testes, defects to the other side and suppresses testes development. In such cases the XY karyotype, complete with its SRY gene, develops a female phenotype (Allen 2007).

When the testes are fully developed they begin producing androgens, which activate androgen receptors in the brain. Androgen is then converted to estradiol, the major estrogen, by the enzyme aromatase to masculinize the brain.[1] This brain sexing takes place during the second half of gestation, and as a result "the structure and functioning of these regions become altered, as are the behaviors they control . . . high concentrations of prenatal androgens result in male-typical behavior . . . female-typical behavior develops in the absence of androgens" (Yang, Baskin, & DiSandro 2010:154). The testes also produce Mullerian-inhibiting substance (MIS) that causes internal female sex organs to atrophy (Swaab 2004).

Brain masculinization is not an all-or-nothing process. Rather, it is one that describes a continuum that may contain significant XX/XY overlap. To grossly simplify, at this level of analysis, sex/gender may be viewed as a continuum from extreme femaleness (which we can define as the complete absence of androgens or insensitivity to them) to extreme maleness (which we can define as high average levels of androgens). The female fetus is protected from the diverting effects of androgen, but not completely. Once prenatal androgens have sensitized receptors in the male brain to their effects, there is a second surge from about the second week of life to about the sixth month of life that further imprints the male brain, followed by the third surge at puberty that activates the brain circuits organized prenatally to engage in male-typical behavior (Sisk & Zehr 2005). All the additional steps required to switch the male brain from its default female form is the reason that significantly more males than females suffer from all kinds of neurological problems (ADHD, dyslexia, autism, Asperger's syndrome, stuttering, language delays, and so forth); things can go awry when perfectly good systems are meddled with.[2]

Disorders of Sex Development: What Can They Tell Us about Gender?

The process of sexing the brain is a real wonder of nature: consider the immensely complicated interactions and permutations of chromosomal, genetic, enzymatic, and hormonal factors that go into it. In the vast majority of cases, the process conforms to the Goldilocks principle, in which everything falls within its allotted margins and goes "just right." But sometimes Goldilocks finds that the bed is too hard, the chair too tall, and the porridge too cold, and she doesn't get what the chromosomes led her to expect. These are individuals born with a number of congenital conditions in which chromosomal/gonadal sex does not necessarily comport with gender identity or with anatomical sex. Today, these experiments of nature are called disorders of sex development (DSDs); in the past, individuals with DSDs were called pseudohermaphrodites or intersex anomalies. Fausto-Sterling (2002) tells us that about 1.7% of individuals are born with some sort of DSD.

DSD individuals are useful to behavioral scientists who want to get to the bottom of gender differences because they defy our usual binary gender categories and expectations. Among non-DSD individuals, gender socialization is strongly confounded by biological sex, because laypersons know (even if constructionists don't) that the gender of their offspring naturally maps to their biological sex. Because gender maps so closely to sex, it is difficult to untangle the relative contributions of biology and socialization to gender-typical personality and behavior. Kenneth Zucker (2002:6) points out that this difficulty has led a number of researchers "to the study of children with intersex conditions in the hope of providing at least a partial solution to the problem." Using the anomalous to gain insight into the normal is a well-worn backdoor approach in science. Species mutants serve biologists to clarify the species norm, and brain-damaged patients provide neuroscientists with a wealth of information about the function of undamaged brains. Likewise, the study of DSD individuals can provide behavioral scientists with important clues about the extent to which gendered behavior rests on biological sex by examining the effects of prenatal hormones (or their absence) on the personalities and behaviors of these individuals (Gooran 2006).

At the extreme feminine end of the feminine-masculine continuum are *androgen insensitivity syndrome* (AIS) individuals. If there were such creatures as scientific essentialists (as essentialism is understood by social constructionists), the AIS phenotype would be a great place for constructionists to engage them. AIS individuals have all the per se "essentials" of maleness, such as an XY karyotype and testicles, but they are female in mind and body and have been considered as such by themselves and others since birth. This anomaly exists because the receptor sites that normally bind androgens are partially (PAIS) or completely (CAIS) inoperative due to a mutation of the androgen receptor gene located on the X chromosome. If the receptors are completely inoperative, the XY genotype develops a female phenotype.

CAIS individuals have the SRY gene and thus have androgen-producing testes (undescended), but because their androgen receptors are insensitive to its effects, the internal male sex structures do not develop. Neither do CAIS individuals have female internal sex organs, because their testes secrete normal amounts of MIS, which atrophies them. The external genitalia are unambiguously female, although the vagina is shallow and leads to a dead end. This condition is typically not diagnosed until the teen years, when AIS individuals consult a physician about failure to menstruate or about painful intercourse. Unresponsive to the masculinizing effects of androgens on the brain, CAIS individuals tend to conform to typical attitudinal, trait, and behavioral patterns of normal females, often exaggeratedly so. They also remain comfortable with their sexual and gender identities after their condition is revealed to them (Jurgensen et al. 2007).

PAIS individuals are usually behaviorally intermediate in terms of gendered behavior. PAIS occurs less frequently than CAIS, and because they are only partially insensitive to androgens, the ambiguity of their genitalia varies with the degree of androgen resistance. PAIS children are assigned and reared as males or females, mostly according to the degree of genital virilization, because the degree of genital viriliza-tion roughly indexes the degree of brain masculinization. Most PAIS individuals report satisfaction with their assigned sex/gender identity (Byne 2006), although there is a small percentage (11 to 14%) who express anxiety, confusion, or discomfort about their assigned gender (gender dysphoria) and want to change it, either from male to female or female to male (Cohen-Kettenis 2010).

Because it is the most common DSD, the most extensive studied condition is *congenital adrenal hyperplasia* (CAH). Classic CAH (about 95% of all cases) is caused by a deficiency in 21-hydroxlase, an enzyme that helps to convert progesterone to cortisol. The deficiency leads to a buildup in progesterone, which is a precursor of testos-terone (T), resulting in high levels of interuterine T and low levels of cortisol (Meyer-Bahlburg et al. 2006). This results in precocious sexual development in males and variable degrees of masculinization of the genitalia and brains of females. The degree of masculinization of the genitalia (a penis-like clitoris and some degree of fusion of the labia) indexes the degree to which the brain has been masculinized. However, because even the most virilized of girls have internal female organs capable of reproduction, most authorities recommend female assignment via hormonal treatment and surgery, despite the elevated risk of such girls later rejecting that assignment (Byne 2006).

CAH females engage in significantly more male-typical behavior and possess more male-typical traits than non-CAH females, such as a liking for rough-and-tumble play, better visuospatial than verbal skills, lower maternal interests, less interest in marriage, a greater interest in careers, and a greater probability of bisexuality and homosexuality, with about one-third of them so self-described (Gooren 2006). Not only do they score higher on more male-typical skills than their unaffected sisters, they also tend to dislike feminine frills such as jewelry and makeup and playing with dolls, suggesting that gender toy preferences are not arbitrary (Garrett 2009). Although there are no studies directly assessing antisocial behavior among CAH women, because they score higher than non-CAH women on traits positively associated with antisocial behavior (aggressiveness, risk taking), and lower on traits negatively associated with antisocial behavior (maternal interest, commitments to relationships, and lower empathy scores), they are likely to be present in female delinquent and criminal populations in numbers relatively greater than their numbers in the general population (Hines 2011).

There are less dramatic but significant behavioral consequences caused by prenatal exposure to androgens among females not considered DSDs. Female fetuses exposed to androgenizing drugs such as diethylstilbestrol (DES) show masculine behavioral patterns as girls and women (Garrett 2009) and are markedly more likely to be lesbian or bisexual than non-exposed females (Gooren 2006). Taken to prevent miscarriages, DES was removed from the market after its effects became apparent. There are also some masculinizing effects noted for females who shared a womb with a male co-twin (Craig, Harper, & Loat 2004), and a longitudinal cohort study by Hines and her colleagues (2002) reported a positive linear relationship between fetal T collected from amniotic fluid and degree of masculinized gender behavior in young girls measured by toy, playmate, and activity preferences. The extent to which prenatally androgen-exposed females move toward typical male preferences most likely depends on the extent of that exposure (Hines 2006).

At the maleness extreme of the continuum *are XYY individuals.* XYY syndrome is the anomaly that has generated more interest than any other among those interested in antisocial behavior (Briken et al. 2006). XYY males are not supermales or born criminals, as used to be thought, but they evidence exaggeration of male-typical behavioral traits. Most descriptions of the behavioral phenotype suggest that compared to XY males, XYY males have higher levels of aggression, hyperactivity, and impulsiveness, an unbalanced intellectual profile (a performance IQ significantly greater than verbal IQ), and atypical brain-wave patterns (Briken et al. 2006; O'Brien 2006). Plasma testosterone concentrations of XYY men are usually found to be at high average levels. Although most XYY males lead fairly normal lives, they are at elevated risk for a diagnosis of psychopathy and for committing sex crimes, and they are imprisoned or in psychiatric hospitals at rates greatly exceeding their incidence in the general population (Briken et al. 2006).

Another interesting intersex condition is an enzymatic condition known as *5-alpha-reductase deficiency* (5-ARD). Because of this deficiency, T cannot be converted to dihydrotestosterone (DHT), which is the androgen required for normal masculinization of the external male genitalia. Thus, 5-ARD males are born with ambiguous or completely female genitalia and almost always reared as girls. However, at puberty T rather than DHT is responsible for the emergence of male secondary sexual characteristics and penile growth, and in 5-ARD males the testes descend and the clitoris markedly enlarges to become a penis, which varies considerably in the degree of virilization (size). In one study of 18 5-ARD boys who had been reared as girls from birth, 17 changed to a male gender identity at puberty (Diamond 1999). There are a number of other studies of 5-ARD individuals that report around a 90% gender switch at puberty (see Byne 2006), although in a study of 25 Brazilian female-raised 5-ARD males, only 13 (52%) changed to male identity at puberty (Mendonca et al. 2003). The varying proportions of gender changing may reflect different cultural attitudes regarding gender and/or varying levels of androgen imprinting (Diamond 2009).

On the other hand, individuals with a condition called *17beta-hydroxysteroid dehydrogenase deficiency* (17-BHSD) show less dramatic gender changes at puberty. These individuals are also XY karyotypes born with ambiguous or fully female genitalia and are typically reared as girls. At puberty they develop similarly to 5-ARD boys, but only about 50% switch gender identity (Gooren 2006). The different rate of gender switching between 5-ARD and 17-BHSD individuals is probably a function of the roles of the two different enzymes involved. In 5-ARD individuals the problem is conversion of T to DHT, which does not affect T's role in masculinizing the brain. In 17-BHSD individuals, the deficiency is in the enzyme that catalyzes the final step in T synthesis, thus leading to a deficit of T available for brain masculinization in utero (Byne 2006).

Transsexuals

Transsexuals are persons who feel that they inhabit the body of the wrong sex. These individuals reveal the complication of sexing what all mammals are prior to that momentous sixth week—undifferentiated hermaphrodites—because they seem to defy the organizational-activation process we have been discussing. How can a male-to-female (MtF) transsexual with completely virilized genitalia, and thus with all the requisite androgens for further masculinization, know that he is a she? Gender identity is obviously very important to humans if trans-sexuals are willing to endure the pain and financial cost of surgical and other procedures to synchronize their minds and bodies.

Data from 17-BHSD individuals may offer some insight. Because of their normal male genitalia, MtF transsexuals must have retained the ability to convert T to DHT. But genital sexing takes place weeks before brain sexing, and these two processes are independent of one another (Hare et al. 2009). Thus, while the genitals develop the normal XY way, the androgen receptors may have be compromised by 17-BHSD deficiencies leading to the failure of brain masculinization (Byne 2006). A number of studies have shown that there are significant trait differences attributable to brain morphology in MtF transsexuals that existed prior to hormone treatment. These differences are intermediate between females and control males (Diamond, 2009).

Another possibility (not mutually exclusive) is that transsexuals have less efficient androgen receptors (AR). Studies have shown that MtF transsexuals have a significantly greater percentage of the long repeat polymorphism of the AR gene than control subjects (e.g., Hare et al. 2009). A polymorphism is a variant (allele) of a gene coding for a protein—in this case the AR receptor. A long repeat means that the gene repeats a nucleotide (the familiar letters of the DNA code) sequence more times than the short repeat. The upshot is that the long repeat of the AR gene leads to less efficient T signaling, and thus to less brain masculinization relative to the short repeat versions.

Further indicative of the prenatal role of androgens are studies that show MtF transsexuals and individuals with gender identity disorder (e.g., Kraemer et al. 2009) have high 2nd to 4th finger length ratios (2D:4D). Females tend to have even ratios, and males tend to have low ratios; i.e., second (index) finger shorter than fourth (ring) finger, with MtF transsexuals and gender identity disorder XYs being intermediate. A number of lines of evidence suggest that the 2D:4D ratio reflects (albeit weakly) the degree of brain androgenization. MtF transsexuals may thus not have had intersex bodies at birth to confound physicians, but rather they have intersex brains that will later lead to gender dysphoria (Diamond 2009).

Other Gender-Confounding Conditions

Chromosomally male infants who lack a penis for one reason or another are perhaps even more interesting than DSD persons in terms of what they can tell us about the effects of prenatal hormones on gendered personality and behavior. These conditions are ablatio penis (traumatic loss of penis), cloacal exstrophy (a severe birth defect wherein the bladder and intestines are turned inside out and exposed and the penis split, absent, or severely deformed), and penile agenesis (born without a penis). Males with these conditions have brains that have presumably been masculinized, but because of their abnormal penis status may be surgically assigned as females and reared as such. Male to female sex reassignment is the norm because of the impossibility of constructing a functioning penis; it's easier to dig a hole than to build a pole.

Ablatio penis is best illustrated by the well-publicized case of David Reimer (popularly known as the John/Joan case) who had his penis mutilated in a botched circumcision. In conformity with the extreme environmental ideas about gender in the 1970s, surgeons castrated David and fashioned a vagina for him. He was given estrogen injections and reassigned as a female with the name Brenda. John Money, the major proponent of gender neutrality at birth in the 1970s, assured Brenda's parents that she would become a well-adjusted woman. All surgical, hormonal, and socialization efforts to turn David into Brenda failed dramatically in every respect. When he learned of his medical history, he expressed relief and underwent further surgery to construct a penis (nonfunctioning). The whole tragic story (David committed suicide in 2004) is told in Colapinto (2006). Because of this and five other similar cases of assigned gender rejection cataloged by Diamond (1999), Money came to reluctantly reject the notion of gender neutrality at birth: "Clearly, the brain holds the secrets of the etiology of gender identity differentiation" (Money 1986:235).

Conversely, Bradley, Oliver, Chernick, and Zucker (1998) report on another infant male reassigned as a female due to loss of his penis during circumcision and raised as a female who was still living as a woman at age 28. This individual reported comfort with the assigned gender but always worked at "male" occupations, had male-typical interests, considered herself bisexual, reported that her sexual fantasies were all of females, and was living in a "lesbian" relationship at the time of interview. Thus, self-defined gender can be inconsistent with gender behavior and sexual orientation.

A review of 50 cloacal exstrophy patients found that over half of the XY female-raised patients displayed male behavior patterns and questioned their gender identity, and all displayed interests and attitudes typical of males despite early castration to avoid the neonatal T surge (Woo, Thomas, & Brock 2009). According to one review, about half of male individuals with penile ablation, penile agenesis, or cloacal exstrophy maintain their assigned gender identities, although this is usually assessed in childhood before the third activational surge of T at puberty; there is an increased probability of rejecting the assigned identity as individuals age (Meyer-Bahlburg 2005). Byne (2006) reports on one XY cloacal exstrophy patient raised as a female who underwent a complete sex/gender change at age 52 only after both his parents had died.

The fact that most female-reared XY karyotypes change to a male identity in spite of incongruent genitalia, exogenous estrogens administered to facilitate female physical appearance, and feminine gender socialization represents the triumph of a virilized brain in what must be a series of stupendous psychologically distressing battles. It is not really that surprising that some would choose to retain their assigned gender in the face of internal and external pressures to do so. After all, these individuals are presented with a fait accompli in that they have visibly female sex organs, they have been treated all their lives as females and are known as such by their friends and acquaintances, and they realize that there is no way that surgeons can fashion a functioning penis for them. One the other hand, those who maintain their gender of rearing offer some evidence that gender identity is not completely or overwhelmingly determined by biology. It is possible, however, that if adequate androgen imprinting of the brain did not occur, then sex of rearing could be the dominant factor in determining gender identity.

Conclusion

William Reiner, a major figure in DSD research, dismisses completely the idea that humans are psychosexually neutral at birth, whose gender identities are molded simply by socialization. He states that "The trendy notion that *Homo sapiens* must develop gender identity or any attribute in a divergent mechanism from other primates or even other mammals is species-narcissistic" (in Diamond 2009:625). I would indeed be truly bizarre to claim that, alone among animal species, humans have somehow managed to become independence from their evolutionary history. Just as stone carvers follow the natural forms within the material with which they work, socializing agents intuitively realize that the stone of nature will always determine the general direction their human sculptures will take. No one dismisses the power of socialization to mold biological material *within* each sex-gender. There are certainly enough Daphne Daredevils and Freddie Fearfuls in the world to convince us that there is much gender overlap in personality, traits, abilities, and behaviors. Within-gender variations should not blind us, though, to the large *between*-gender variation consistently found for core gender traits. Indeed, these differences should be expected on the basis of evolutionary logic.

In light of all the prenatal processes experienced by the zygote/embryo/fetus, the amazing prescience of Samuel Taylor Coleridge (he of the Aristotle-Plato divide) is greatly to be admired. Long before genes, chromosomes, or hormones were heard of, Coleridge wrote, "The history of man for the nine months preceding his birth would, probably, be far more interesting and contain events of greater moment than for all the three score and ten years that follow it" (cited in Hepper 2005:474). Indeed, it is this nine-month intrauterine environment that places us on a gender trajectory that socialization, postnatal hormone therapy, or surgery cannot derail—nature plants gender, and nurture cultivates it, but nurture does not, and cannot, plant it.

Endnotes

1. It seems strange to talk about estrogen, which is responsible for the development of female secondary sexual characteristics at puberty, playing a part in sexing the male brain—Mother Nature loves to confuse us. It turns out that during fetal development the female ovaries have not started pumping out estrogens yet, but the male testes do, and estrogen's function is to prevent the premature death of sperm cells. Estradiol, a major estrogen, is a metabolite of testosterone produced by the enzyme aromatase. Evidently, testosterone converting neurons (those containing aromatase and AR) create the male neural circuitry, but their ability to do so depends on the "female" hormone, estrogen (Wu et al. 2009).

2. This is an extremely simplified account of human sexual differentiation. The SRY gene is a regulator or switch gene that turns on a number of other genes. It is only activated prenatally for a short time at various intervals, itself being regulated by autosomal genes that it regulates in a reciprocal feedback way. It is the downstream autosomal genes that go about the

93

business of actually constructing the testes. Fourteen different genes, including two on the X chromosome (no others on the Y) have been identified as contributing to mammalian sex determination. A mutation of one of them on chromosome 11, known as the WT1 gene, leads to yet another intersex condition known as Denys-Drash Syndrome (DDS). DDS results in an XY karyotype with ambiguous or completely feminized genitalia, although behavior is male biased. There are even cases of XX karyotypes that develop male phenotypes due to the SRY gene crossing over from the Y to the X during spermatogenesis (Rosario 2009).

Chapter 8

A Further Journey into the Gendered Brain

Sex on the Brain

Gender is a conceptual schema we have in our heads regarding what it means to be a male or a female. The social constructionist position is that gender is drummed into our heads by the expectations of our socializing agents and by our social roles, because the infant brain is *tabula rasa* onto which society can inscribe anything. Gender is certainly in our heads, but neuroscience tells us that the *tabula rasa* assumption is fundamentally impossible (Tooby & Cosmides 2002). Masculinity/femininity is planted in the brain's structure and function by eons of evolutionary selection pressures and then cultivated in different ways by social expectations, but rarely does it stray too far beyond nature's boundaries. As we have seen, neuroscientists have made great strides in mapping numerous structural and functional differences in the brains of males and females that throw light on gender differences in personality traits, abilities, and behavior.

Journeying into the gendered brain has been described as "neurosexism" by Cordelia Fine (2010) in her book *Delusions of Gender*. Fine's book is an example of my sponge metaphor for social constructionism, because it goads readers into questioning the methods and techniques we rely on to study gender differences. She tells us how modest and even weak findings are dressed up and greatly over-interpreted by the media in ways that support sexist stereotypes. Fine places the blame primarily on the media, but she also blames neuroscientists themselves. Popular books such as the enormously successful *Men Are from Mars, Women Are from Venus* do a disservice to serious science, and Fine wants to remind us that men and women are from the same planet.

Fine strikes me as an orthodox constructionist feminist with a yearning for global unisex. She has uncovered some shoddy research in the neuroscience of sex-gender differences and deduced that all such science is bad science. Fines' work is a much more sophisticated version of the works of the feminists who wanted to fry Bacon for his sexist metaphors in Chapter 3, but it is still overstated. We hear the same complaints of determinism, reductionism, and essentialism, with the added grumble that neuroscience assumes that female brains are hardwired for their supposedly inferior roles. Yet Fine makes extensive use of neurological data when it suits her. On the basis of such data on psychopaths, such as their relative inability to tie together the rational and emotional structures of their brains, Fine and Kennett (2004) concluded that society should treat rather than punish them. Neuroscience data are impeccable when they advance a cause she values, but when they do not, they are hopelessly flawed with their implications of essentialism and hardwiring.

Softwiring the Brain

Far from advocating a hardwired view of the brain, brain plasticity has been the guiding principle of neuroscience for at least half a century. Although about 60% of the human genome is involved in brain development (Mitchell 2007), there are too few genes to wire the billions of neurons and the trillions of connections they can make with one another in predetermined ways. Our experiences, not our genes, will largely specify the connection patterns of our neurons. If genes were the only determinants specifying neuronal connections, we would be hardwired drones incapable of adapting to novel situations. Because human environments are so varied and complex, natural selection has favored brain plasticity over fixity. But brain plasticity does not permit anything. Just as the flow of a river is biased by the topography it encounters on its journey to the sea, brain plasticity is biased in sexually dimorphic ways by the topography of the brain carved while in the womb.

Every member of a species inherits identical brain structures and functions produced by a common pool of genetic material, but individuals will vary in brain functioning as their genes interact with their environments to softwire them (Gunnar & Quevedo 2007). Neuroscientists distinguish between two brain developmental processes that physically capture environmental events: experience-expected and experience-dependent (Schon & Silven 2007). Experience-expected mechanisms reflect the brain's phylogenic history and are hardwired, although they require specific experiences to trigger them at critical periods. Experience-dependent mechanisms reflect the brain's ontogenic plasticity. The distinction between the two processes is illustrated by language, the *capacity* for which is an entirely hardwired experience-expected capacity, but what language(s) a person speaks is entirely the result of softwired experience-dependent processes.

Certain abilities and processes such as sight, speech, depth perception, affectionate bonds, mobility, and sexual maturation are vital, and natural selection has provided for mechanisms (adaptations) designed to take advantage of experiences occurring naturally within the normal range of human environments. These processes that have evolved as a readiness of the brain at certain critical developmental periods to assimilate into its pathways information that is vital to an organism and ubiquitous in its environment. Some things are so important that they cannot be left to the vicissitudes of learning, so our brains are organized to frame and orient our experiences so that we will respond consistently and stereotypically to vital stimuli (Geary 2005).

Experience-dependent brain development relies on experience acquired during the organism's development, which includes gender socialization. Much of the variability in the wiring patterns of the brains of different individuals depends on the kinds of physical, social, and cultural environments they will encounter. It is not an exaggeration to say that "experience dependent processes are central to understanding personality as a dynamic developmental construct that involves the collaboration of genetic and environmental influences across the lifespan" (Depue & Collins 1999:507). Although brain plasticity is greatest in infancy and early childhood, a certain degree is maintained across the lifespan so that every time we experience or learn something, we shape and reshape the nervous system in ways that could never have been genetically programmed. There are certainly arguments in neuroscience about brain developmental processes, but they are not about "*whether* the environment thoroughly influences brain development, but *how* it does" (Quartz & Segnowski 1997:579). This central tenet of neuroscience hardly sounds like the hardwired determinism that supposedly infects it.

Brain Laterality

The most recent evolutionary addition to the brain is the cerebrum, which forms the bulk of the human brain. The cerebrum is divided into two complementary hemispheres that are connected by the corpus callosum. It is generally accepted that the right hemisphere is specialized for perception, motor skills, spatial abilities, and the expression of emotion, and that the left hemisphere is specialized for language and analytical thinking (Parsons & Osherson 2001). Although the hemispheres have specialized functions, they work in unison like two lumberjacks attacking a tree at opposite ends of a saw. Female brains are less lateralized than male brains, which implies higher functional connectivity in female brains, that their cerebral hemispheres are less devoted to specialized tasks, and that both hemispheres contribute more equally to similar tasks than they do in males (Luders et al. 2006).

Testosterone (T) is strongly implicated in the process of lateralizing the brain. Chura and her colleagues (2010) find that increasing amounts of fetal T is significantly related to increasing rightward asymmetry of the corpus callosum in males. Fetal T slows down the maturation of the left hemisphere in the male brain, thus allowing the right hemisphere to gain dominance, which explains the enhanced performance of males in right hemisphere–related visuospatial tasks. The more symmetrical female brain means that there is more active cooperation between the hemispheres, which leads to better synchronization of emotional and cognitive processing. Neuroimaging studies show that women can more readily access and assess the rational and emotional content of social messages simultaneously, as indexed by observed blood oxygen flow across the hemispheres (Lippa 2003). It is known that the higher brain regions of the cerebrum (its outer layer, the cortex) develop sooner and faster in females, which explains accelerated language development in females.

Hemispheric specialization is nicely illustrated in patterns of brain activity while completing IQ items simple enough that almost anyone can complete them. It is well established in neuroscience that higher IQ individuals have more efficient brains. Neural efficiency is best tested using positron emission tomography (PET) scans that measure cerebral glucose metabolism as the brain takes up positron-emitting glucose administered to subjects by injection or inhalation. A computer reveals colorized biochemical maps of the brain identifying the parts activated while engaged in some task as the energy supplied by the glucose is metabolized. The more difficult an item is for a person, the more brain energy he or she has to use to solve it. Glucose metabolic rates at various brain slices (brain levels) are highly inversely correlated with IQ scores, which means that higher IQ persons expend less energy when performing intellectual tasks and possess brains that are speedier, more accurate, and more energy efficient than low IQ subjects (Gray & Thompson 2004).

This inverse neural activation-intelligence relationship is moderated by gender. It holds for males when performing spatial tasks such as figure rotations and for females when performing verbal matching tasks and when identifying emotions in photographs (Jausovec & Jausovec 2008). That is, higher IQ males expend less cerebral energy than lower IQ males when performing spatial tasks, and higher IQ females expend less cerebral energy than lower IQ females when performing verbal matching and emotional tasks. This phenomenon is interpreted as males having more efficient brains for visuospatial tasks and females having more efficient brains for verbal and emotional tasks (Neubauer & Fink 2009).

Male superiority in visuospatial tasks and female superiority in verbal and object location memory is found in every culture where it has been tested. Silverman, Choi, and Peters (2007) examined data from almost 250,000 subjects in 40 countries and found this to be the case. They attribute these universal sex-differentiated brain patterns to the sexual division of labor during the Pleistocene epoch and beyond, in which males were the primary hunters of meat and females the primary gatherers of plant food. Visuospatial abilities have obvious utility in pursuing and hunting animals and then finding the way home at the conclusion of the hunt. On the other hand, object location memory and verbal skills are most useful in locating edible plants among a diversity of vegetation arrangements, remembering that location, and communicating it to others.

These traits are a part of each person's repertoire of cognitive skills measured by IQ tests. A person's IQ score is the sum of two IQ scales—verbal (VIQ) and performance (PIQ) scales—divided by two. Given the gender difference in verbal and visuospatial traits, we should expect females to score higher on VIQ and males to score higher on PIQ. The original Wechsler IQ tests did show highly significant sex differences between the subscales in the expected directions (females significantly greater VIQ; males significantly greater PIQ), but Wechsler wanted a sex-neutral measure of intelligence, not a measure of verbal and visuospatial abilities. He achieved this by pruning the items most responsible for the sex difference; thus, today's items index verbal or performance skills less strongly than did the original items (Wells 1980).

Arousal Levels

Reflective of evolutionary pressures directed at roles for child care and food gathering, females are more mindful of environmental stimuli—what is going on around them. This is demonstrated in

99

numerous studies of memory of spatial configurations in which females consistently outperform males (Silverman, Choi, & Peters 2007). We should not confuse visual configuration abilities, in which females excel, with visuospatial abilities, in which males excel. Visual configuration ability refers to such things as quickly identifying the form of something as determined by the arrangement of its parts and color, or identifying matching items that enables classification of the thing perceived. Visuospatial ability is the ability to visualize objects in space in one's mind, and how they could be viewed from a different perspective.

Greater female attention to environmental details may reflect greater augmentation capabilities of the reticular activating system (RAS) in women. The RAS is a finger-sized bundle of neurons located at the core of the brain stem and feeds arousal stimuli to the thalamus for distribution throughout the brain. It is a sort of information filtering system that broadly determines consciousness, arousal, and alertness by receiving signals from the environment. If we think of the RAS as a radio receiver, we might say that it is more fine tuned in females and that the sound is turned up. A more alert RAS may partially account for why females are less prone to boredom (Gemminggen, Sullivan, & Pomerantz 2003) and to sensation seeking (Zuckerman 2007). Boredom (under arousal) is an unpleasant condition that motivates seeking more sensory input to alleviate it. This is a major symptom of attention deficit with hyperactivity disorder (ADHD). Deficits in the RAS have been invoked to explain why males are significantly more likely to be diagnosed with ADHD than females (Hermens et al. 2004). Among those with the most serious of the conditions along the ADHD spectrum (attention deficit combined with hyperactivity), 7.3 males are diagnosed for every female (Rhee & Waldman 2004).

Dolls, Trucks, Evolution, and the Visual System

One of the perennial issues surrounding gender socialization is sex–linked toy and color preferences. These preferences have been used to argue both for the innateness of gender and for the power of socialization to mold gender. In Chapter 6, Barrie Thorne (1993:2) argued that toy preferences showed that gender is socially constructed via arbitrary societal norms: "Parents dress infant girls in pink and boys in blue, give them gender differentiated names and toys, and expect them to act differently."

Toy preferences inconsistent with genital sex and assigned gender, such as observed by CAH girls, is an embarrassment to social constructionists who put the inconsistency down to atypical (i.e., male-typical) gender socialization. The argument is that because of the virilized genitalia of these girls, their parents treat them more like boys (Fausto-Sterling 2002), but clearly common sense should dictate exactly the opposite. This is exactly what the research shows; because of their male-like behavior, their parents strive extra hard to socialize them in gender-appropriate ways (Hines 2011; Pasterski et al. 2005). We see these sex differences in toy preferences in infants and toddlers when they are allowed to choose, and among non-human primates as well (Hassett, Siebert & Wallen 2008; Hines & Alexander 2008). Neither of these observations can reasonably be attributed to socialization. The contemporary evidence points strongly to the conclusion that sexually dimorphic toy preferences reflect basic neurobiological differences between males and females that ultimately reflect evolutionary logic.

It seems that the origin of apparently innate biases for gender-typed toy preferences may be based on the influence of androgens on the visual pathways from the retina to image processing centers of the brain. There are a number of different types of cells that send visual information from the retina to the brain, with the two most pertinent ones being parvocellular (P-cells) and magnocellular (M-cells). P-cells (the "What is this thing?" cells) transmit information about the color and shape of stationary objects, and M-cells (the "Where is this thing in space?" cells) carry information about depth and motion. Research consistently shows that females have significantly greater density of P-cells and males have significantly greater density of M-cells, which is consistent with the superior skill in the visual configuration ability to identify shape and color among females, and with the superior visuospatial skills in seeing motion and depth in males (Alexander, Wilcox, & Woods 2009).[1]

The evolutionary link is easy to discern here. Female gatherers needed to recognize immobile plants by their shape and color (what it is) while male hunters needed to process the motion of prey or predator (where it is) to make a successful kill and to avoid being the dinner rather than the diner. Natural selection supplied the mechanisms to allow our ancestors to better perform their roles. These evolutionary mechanisms are reflected in today's gender-differentiated infant/child toy preferences. Boys' preference for moving objects such as toy trucks and balls is biased by their perceptual M-cells because these objects move in space and can be manipulated. Girls' preferences for dolls provide them opportunities to practice nurturance, and being more drawn to faces than moving objects is biased by the P-cell advantage (Alexander 2003).

Some may argue that evolutionary explanations for gender-based color and toy preferences are "just so" stories. But there has to be some ultimate-level explanation for why P-cell/M-cell sexual dimorphism is present in contemporary humans, as well as many other gender differences in the visual cortex that bias color preference (Amunts et al. 2007). It is common practice in biology to inquire into the fitness functions of any morphological, physiological, or behavioral trait they observe in any species. No one has come up with an alternative explanation for the sexual dimorphism in perceptual differences, which once again stresses how we can ill afford to ignore evolutionary theory in our thinking about gender.

Sugar and Spice; Snips and Snails

Robert Southey, the author of the 19th-century metaphorical nursery rhyme telling us what girls and boys were made of, had something to say about the respective natures of males and females. I suspect that Southey was impressed by the huge gap between the sexes in pro- and anti-social behavior, especially criminal behavior, that others have noted in all cultures at all times. Writing about what I considered the bare minimum for explaining the universal sex difference in criminal behavior, I concluded that it would have to be sex differences in empathy and fear (Walsh 2011:124):

> Empathy and fear are the natural enemies of crime for fairly obvious reasons. Empathy is other oriented and prevents one from committing acts injurious to others because one has an emotional and cognitive investment in their well-being. Fear is self-oriented and prevents one from committing acts injurious to others out of fear of the consequences to one's self.

Sex-differences in empathy and fear (females higher on both) evolved in response to sex-differentiated reproductive roles. Empathy assured the rapid identification and provision of infant needs and nourished social relationships (de Waal 2008). Fear kept both mother and child out of harm's way and provided a sturdy scaffold around which to build a conscience (Campbell 1999). Many other prosocial tendencies flow from these two basic foundations, such as altruism, self-control, guilt proneness, and agreeableness. While these traits contain a heavy dose of socialization, they are all strongly related to differential brain functioning (and that in turn by gene functioning), which makes socialization either easier or more difficult to take hold.

Empathy

Empathy is an ancient phylogenic capacity that evolved rapidly in the context of mammalian parental care (de Waal 2008). Empathy is an integral component of the love and nurturing of offspring, because caregivers must relate quickly and automatically to the distress signals of their offspring. Offspring care in all mammalian species is primarily maternal, thus selection pressures for empathy would have operated more strongly in females. Mothers who were not alerted by their offspring's distress signals or by their smiles and cooing are surely not among our ancestors.

Greater female empathy may be traced to the effects of low testosterone (T) and/or to higher oxytocin (OT) functioning in females, because T and OT are mutually antagonistic (Knickmeyer et al. 2006). Using functional magnetic resonance imaging (MRI), studies have shown that females who receive a single sublingual dose of T show a significant reduction in empathetic responses (Hermans, Putman, & van Honk 2006), and males given a single intranasal dose of OT significantly enhanced their ability to infer the mental states of others (Domes et al. 2007). Thus, males become more empathetic with OT, and females become less so when given T. These responses take place outside conscious awareness, because the target sites for both T and OT are located in the "emotional brain"—the limbic system. An fMRI study comparing neural correlates of empathy found that females recruit far more emotion-related brain areas than males when processing empathy-related stimuli. Males tended to recruit brain areas associated with cognitive evaluation rather than emotional evaluation (Derntl et al. 2010).

The neural architecture that gives rise to empathy is considered to reside in so-called mirror neurons. Mirror neurons are brain cells that fire (respond) equally whether an actor performs an action or witnesses someone else performing the action. Thus, outside the observer's conscious awareness, the observer's neurons mirror the behavior of another as though the observer were acting in the same way. This unconscious communication between the neurons of one person and those of another reflects a correspondence between self and other that turns an observation into empathy. Studies using fMRI have shown that those subjects with higher empathy scores on a variety of empathy scales show stronger brain activation to empathy-evoking stimuli (Schulte-Ruther et al. 2007).

As we expect from the logic of natural selection, a number of studies of the human mirror neuron system find that females tend to be better than males in mirroring the emotions of others. One fMRI study concluded that "females recruit areas containing mirror neurons to a higher degree than males during both self- and other-related processing in empathetic face-to-face interactions" (Schulte-Ruther et al. 2008:393). Other evidence comes from EEG studies of brain wave activity showing that females (and high-scoring males also) tend to have patterns of rhythm activity indicating a high level of mirror neuron activation (Cheng et al. 2008).

Fear

The other half of the equation is fear. Fear is a basic affective state that signals danger. It is an unpleasant state of arousal that motivates an organism to escape the immediate threat and to avoid being in similar positions in the future. Fear is thus adaptive in that it facilitates the emergence of escape/avoidance behaviors that enhance an organism's chances of survival and reproductive success. Anne Campbell's (1999) "staying alive" hypothesis proposes that sex differences in criminal behavior are based ultimately on parental investment. Because the obligatory parental investment of females is enormously greater than that of males, and because of the infant's dependence on its mother, a mother's presence is more critical to offspring survival (and hence to the mother's reproductive success) than is a father's. There are no human cultures in which mothers desert their children anywhere near the rate of fathers (Campbell, Muncer, & Bible 2001). Unlike males, females are limited in the number of children they can have, so each child represents an enormous personal investment that they will not relinquish without the most compelling reasons to do so. Thus, the reproductive success of females lies primarily in parenting rather than mating effort, and this requires staying alive.

Because a female's survival is more critical to her reproductive success than is a male's, Campbell argues that females have evolved a stronger tendency than males to avoid engaging in behaviors that pose survival risks. The practice of keeping nursing children in close proximity in ancestral environments posed an elevated risk to both mother and child if the mother placed herself in risky situations. The evolved proximate mechanism to avoid doing so is a greater propensity for females to experience more situations as fearful than do males. Fear of injury accounts for the greater tendency of females to avoid or remove themselves from potentially violent situations and to employ indirect and low-risk strategies in competition and dispute resolution relative to males.

Females experience fear more readily and more strongly than males, whether assessed in early childhood (Kochanska & Knaack 2003), the middle-school years (Terranova, Morris, & Boxer 2007), or among adults across a variety of cultures (Brebner 2003). A meta-analysis of 150 risk experiment studies found that sex differences were greater when the risk involved meant actually carrying out a behavioral response rather than simply responding to hypothetical scenarios requiring only cognitive appraisals of possible risk (Brynes et al. 1999).

A part of the brain's limbic system called the amygdala is crucially involved with processing fear. Neuroimaging studies have shown sex-related hemispheric laterality of the amygdala with males specializing to the right amygdala (specializes in detecting salient emotional stimuli) and females to the left (involved in sustained stimulus evaluation) (Cahill et al 2004; Williams et al. 2005). The frontal cortices also play a crucial role in modulating impulsive behavior initiated by the amygdala, and females have been found to have a highly significant greater ratio of orbital frontal cortex volume to amygdala volume. This suggests that females are less likely to express negative emotions in aggressive ways (which could lead to injury) and to internalize stressful emotional experiences instead (Gur et al. 2002). It also may explain why females have, without exception, showed greater levels of constraint/self-control across numerous studies regardless of differences in data, methods, culture, or ages of subjects (Chapple & Johnson 2007). Self-control is a major correlate of criminal and analogous behaviors (Pratt & Cullen 2000).

Conclusion

Despite claims of neurosexism, neuroscience is the only way that we can gain hard, tangible evidence about the fundamental reasons for gender differences in traits, abilities, and behavior. Why it is derogatory rather than celebratory to talk about sex differences is beyond me. It is difficult to believe that the vast majority of human males with their common-sense understanding of sex differences and with mothers, daughters, sisters, and wives would harbor attitudes that could be harmful to them. Surely the stereotypes we have about males are more unflattering (aggressive, violent, domineering, insensitive, and so forth) even if they are true *as generalities*. Evolution has molded males and females to carry out the *only* imperative that nature has—the continuance of life—which is achieved via individual survival and reproductive success. It falls upon the shoulders of females to perform the most important and noble role of all: that of carrying, delivering, and nurturing the next generation of humans. That is both a burden and a crowning glory, assigned not by a patriarchal society but by Mother Nature herself. Observing a young mother basking in her newborn, we see the quintessence of joyous satisfaction, certainly a far cry from an inferior role, as it is often described by radical feminists. To liken traditional sex roles to slavery and prostitution, and all heterosexual sexual activity to rape, as many of the most radical feminists have done, is the essence of idiocy and bigotry, and it is not too far away from self-loathing. Furthermore, it is an insult of the most egregious kind to millions of men and women living decent, moral lives in the context of those traditional roles. My wife hardly thinks of herself as a slave, nor do I think of her as a prostitute.

Endnote

1.   Alexander, Wilcox, and Woods (2009) provide us with a ready example of the non-essentialism of neuroscience. Constructionists jump on findings about gender difference scores as implying that all males have a preference for trucks or all females for dolls. One way of quantifying gender differences on any trait or behavior are effect sizes rendered as a Cohen's d. Because d is expressed in terms of standard deviation units, it is equivalent to a z score corresponding to an area under the normal probability curve. Taking the d for girls' preference for the dolls over trucks, the reported d is 1.27. This represents an area under the curve of .898. The intuitive interpretation is that the average for girls is at about the 90th percentile of the boy's distribution in their preference for dolls over trucks, or that about 10% of the boys score above the average for girls. The corresponding figure for fixation on the truck was that the average boy would be at the 78th percentile of the girl's distribution, or conversely, 22% of girls score above the boy's average. While these differences are wide and statistically significant, they can hardly be taken as essentialist, discrete, non-overlapping groups.

Chapter 9

Race and Racism

Race as Social Construct

If talk of the existence of gender differences rooted in biology is provocative, then talk of race as a biological entity is positively incendiary, and to avoid a minefield is a wise thing. African American sociologist William J. Wilson tells us that social scientists have indeed tiptoed around racial issues or addressed them in "circumspect ways" (1987:22). The only position on race with the seal of approval in social science is that it does not exist as a biological entity, a position underscored by placing the term in scare quotes every time it appears. The objection to the race concept is understandable in light of the tragic overtones of the term, and we should doubtless spike our observations about it with a healthy helping of prudence, but to insist that race must be examined and interpreted only from the constructionist perspective amounts to a professional gag order.

Anti-race scholars believe that the race concept is a social construction that is very real socially but lacks any biological undercoating, that it is a bad thing, and that we should get rid of it. These folks have been trying to kill the word and the concept for almost a century, but every time we think it dead and buried, it returns to life with renewed vigor. Over the years, whole books have been written to strongly deny the biological reality of race (Graves 2001), which have been countered by others just as strongly, asserting its biological reality (Sarich & Miele 2004). Foster (2009:357) points out that these polarized positions are based on two different logics and standards of evidence, complaining that the critics of race publish their missives only in social science journals and "do not engage the scientific arguments."

Race gains and loses its reality from time to time depending on a number of factors both scientific and ideological. The United Nations

Educational, Scientific and Cultural Organization's (UNESCO) 1950 statement on race denied its validity, but the UNESCO committee's 1951 statement defended it. Graves (2001) saw the change as the result of the 1951 committee being composed of more politically conservative members relative to the liberal membership of 1950. Thus, the truth about the existence of race that emerged from these two committees depended more on their respective ideological composition than on data. Both the 1950 and 1951 statements, however, were made in relative ignorance of genetics compared with what we know today.

In addition to ideological swings, the march of scientific technology and knowledge also brings periodic reappraisals of the concept, although it will face stiff competition from constructionists regardless of the quality of the data. Ann Morning's (2008) analysis of biology textbooks published from 1952 to 2002 found a U-shaped pattern of interest in race. There was considerable interest in the 1950s, when race was defined by morphology and still viewed somewhat in terms of superiority/inferiority. There was a virtual disappearance of discussions of race in the 1980s in response to invidious comparisons, and a sharp return in interest in the 1990s as genetic knowledge increased exponentially.

The human genome project has reframed the race concept as geneticists look for evidence for its existence at the molecular level. Some (Duster 2006) argue that this development had led to the FBI's establishment of a criminal DNA data base that can be used against minorities. Others view it positively, such as the announced intention of Howard University to build a database of African American DNA to "jump-start an era of personalized medicine for black Americans" (Kaiser 2003:1485). Evidently, black intellectuals are not as anxious to jettison race as are their white counterparts, but then, blacks are allowed to express racial pride in ways that whites are not. Nadia Abu El-Haj (2007:284) notes some of the many practical uses of race and opines: "Given the prevalence today of race in the practices of biomedicine, pharmacogenomics, forensics, population genetics, and a variety of other genomic and postgenomic fields, it would seem that those scholars who argued for the revalidation of race were correct." We will explore what the genomic sciences have to say about race in Chapter 11; for now, we concentrate on the word and concept of race and its association with racism.

Race in Antiquity and Its Relationship to Slavery

Race is an ambiguous term fraught with historical, philosophical, ethical, political, and scientific disputes. The definitional problems are reminiscent of U.S. Supreme Court Justice Potter's declaration in a pornography case that he could not define obscenity but knew it when he saw it. Likewise, everyone knows race when they see it. The social constructionist complaint is that people tend to assume that physical characteristics used to identify race are correlated with certain personality and behavioral differences, thus placing those who share those physical characteristics in the same boat. Because of this, we are enjoined not to engage in group essentialism by praising one group, such as Asian Americans, as the model minority, lest doing so engage thoughts about other minority groups as less than model (Sayer 1997).

Social constructionists clearly engage the race concept as a normative rather than scientific issue. Race is dismissed as an illusion in the American Anthropological Association's (AAA) *Statement on "Race."* The statement avers that race is "a social mechanism invented during the 18th century to refer to those populations brought together in colonial America: the English and other European settlers, the conquered Indian peoples, and those peoples of Africa brought to provide slave labor" (AAA 1998:1). Thus, race is seen as something conjured up by Europeans and European Americans to justify colonialism and slavery, respectively. This is a position that appeals to motives and not one that confronts modern theories of race on their merits. It is a claim that is repeated constantly but also one that ignores a huge literature pertaining to ideas about race from cultures around the world that existed many centuries before the eighteenth. These cultures did not have the English word *race*, of course, but they had words that described much the same thing, and concepts to match.

According to Banton (2010), the term race first entered the English language in the 16th century, which, with the exception of Spanish incursions into South America, is before any large-scale European slave trade or colonial enterprises commenced. A book-length analysis of the historical origins of the idea of race claims that it began to coalesce in Europe during the 12th century from a variety of sources seeking to explain human differences based on blood, physiognomy, and climate (Hannaford 1996). Still others claim that the ancient Latin term *gens* had a meaning close to the modern meaning of a taxonomical grouping of peoples sorted according to common characteristics and ancestry (Hudson 1996). The 18th-century meaning of race had many extra connotations tacked on to *gens*, but also long before that time Africans had "roughly constituted a single 'race' even in the traditional sense of lineage" (Hudson 1996:249). Hudson offers as evidence the Old Testament story, shared by Christianity, Islam, and Judaism, of Africans being the descendants of Ham, the disfavored and cursed son of Noah.

We do not honestly know if Europeans invented the word and its attendant connotations explicitly to justify their subjugation of other peoples, or if they took an existing term (such as the Spanish *raza*, for lineage) that seemed appropriate for classifying the increasing number of different peoples that they encountered on their voyages of discovery. We do know, however, that Islamic slavery existed long before the 18th century and was never deemed in need of justification, because this foul practice was simply a fact of life. If Muslims ever felt the need to justify slavery, they did so by invoking the curse of Ham. As 10th-century Arab historian Akhbar az-Zaman (in El Hamel 2002:40) wrote, "Traditionalists say that Nuh [Noah], peace upon him, cursed Ham, praying that his face should become ugly and black and that his descendants should become slaves to the progeny of Sam."

Despite statements such as this, many Western writers have come close to idealizing Muslim slavery, telling us that master and slave lived in racial harmony. The image of a man who is owned living in harmony with a man who owns him coheres so poorly with common sense that one wonders how such a notion could have ever materialized. Historian Bernard Lewis supplies us with a plausible answer (1990:101): "The myth of Islamic racial innocence was a Western creation and served a Western purpose. . . . [It] provided a stick with which to chastise Western failings." The countries that did most to end slavery, the United States and Britain, have produced the majority of scholars who reproach their own countries by contrasting the supposed racial harmony of other cultures with their own.[1] Lewis later declares that this self-castigation is but another iteration of Kipling's "white man's burden," only this time it is a burden of guilt because it is an "insistence on responsibility for the world and its ills that is as arrogant and as unjustified as the claims of our imperial ancestors" (1990:102).

Racism East and West

No one needs convincing that we humans are a bigoted lot, for history is replete with efforts to dehumanize one group of people or another along racial, ethnic, national, religious, linguistic, or any other fault line that separates *them* from *us*. There are different forms of bigotry, with racism being a more insidious form than others that expound philosophies of in-group superiority and fear and dislike of out-groups, such as xenophobia and ethnocentrism. These forms of bigotry may be alleviated by contact with, and assimilation of, the out-group, just as religious bigotry is assuaged by conversion. Racial bigotry does not allow assimilation, and conversion is not an option because victims of racism are persecuted for what they are, not because of what they believe. American Jim Crow racism arose from the presence of huge populations of slaves among white Southerners, and then from the specter and subsequent reality of emancipation. Jim Crow racism was a virulent ideology claiming an essentialist inferiority of blacks and was a combination of overt bigotry and legal, political, and social discrimination (Bobo & Kluegel 1997).

The belief among many seems to be that we have racism because we have race: No race, no racism. Anyone who studies race as a reality or is interested in racial differences is considered racist, even if he or she has done nothing other than to report findings contrary to the no-race position, or that show a particular racial group in a poor light relative to other racial groups. Racism is certainly a pernicious disease, but honestly reporting research findings in sensitive areas should never be considered one of its symptoms. It is all too common to conflate the terms race and racism and to tar anyone who studies *race* with the *racism* brush, but as Ruth Benedict wrote long ago, "It is no paradox that a student have on his tongue a hundred racial differences and still not be racist" (1942:vii). The racist label is applied so promiscuously by the self-righteous that all stable meaning has been washed out of it. Such schoolyard name-calling cuts off any meaningful discourse to the advantage of no one, and it has no place in science. All that ad hominem attacks do is acknowledge that one is unable to refute the target's arguments with evidence, although for the anti-science streak in constructionism, evidence doesn't really matter. Regardless of the diverse circumstances under which it is applied, the label can severely burn a career, for it sticks to its victim like hot tar. The wanton use of this rhetorical weapon has resulted in "an unproductive mix of controversy and silence" (Sampson & Wilson 2000:149). Controversy, if engaged in courteously and rationally, can be most productive, but silence is the tactic of the timid and gets us nowhere.

111

Dehumanizing stereotypes supply the scaffolding for any racist ideology directed against any group, but such stereotypes about sub-Saharan Africans have been particularly nasty and numerous. Shelby Steele notes that not only have there been more stereotypes of blacks than of other racial groups, "but these stereotypes are also more dehumanizing, more focused on the most despised human traits: stupidity, laziness, sexual immorality, dirtiness, and so on. In America's racial and ethnic hierarchy, blacks have clearly been relegated to the lowest level—have been burdened with an ambiguous, animalistic humanity" (1991:134).

Contrary to the claims of Western inverse ethnocentrics, such stereotypes are not unique to America or Western Europe. More vicious stereotypes of Africans existed in classical and medieval times, especially in "racially enlightened" Muslim slave societies. Islamic historian Chouki El Hamel (2002:43) informs us that before the coming of Islam in the 7th century AD, the Berbers conquered the blacks of the Sahara and "assumed for themselves a superior status, placing the Blacks in lesser subordinate status. Because the Blacks were different in their cultural and racial characteristics, the racial binary division was easily developed." After the Berbers' conversion to Islam, their racial prejudices "took an Islamic form" (El Hamel (2002:44).

Stereotypes of both black and white (European) slaves existed in Islam, but stereotypes of blacks were far more derogatory. In distinguishing between black and white slaves, the 14th-century Arab historian Ibn Khaldun writes, "Therefore, the Negro nations are, as a rule, submissive to slavery, because [Negroes] have little [that is essentially] human and have attributes that are quite similar to those of dumb animals, as we have stated" (as quoted by Lewis 1990:53). There are numerous examples of the most vile caricatures of blacks in Islamic literature and poetry, including such classics as *The Arabian Nights*. Many of these images, including self-image descriptions attributed to blacks, equal or exceed those found in the Cotton South of the United States.

Kalduhn's idea of natural slaves is similar to Aristotle's notion that the morality of slavery lies in its accordance with nature. By this Aristotle meant that slaves lack deliberation and foresight, which for Aristotle is the per se essence of a human being. He did not deprive slaves of their humanity, however, indicating that under the right master they could gain some semblance of reasoning qualities. But because slaves were considered more like beasts than men, "it is in Nature's design that slaves would be distinguishable *physically* from [their] masters" (Smith 1983:118). The more a slave differed physically from his master, the more likely nature had destined him for slavery, and blacks differed from Greeks more than any other peoples the Greeks enslaved. However, Aristotle counseled kindness toward slaves and believed that given their supposed intellectual deficits, they would benefit from the benevolent care of their masters (Smith 1983). This all sounds very much like the happy darkie stereotype of slaves in Dixie voiced more than 2,000 years before race and racism was supposed to have been invented.

Let us make one last stop in the Middle East before returning to America. There were many blacks in Morocco in the 17th century AD who had either been freed, had run away, or had been abandoned by their masters during harsh times. In 1699, Mawlay Isma'il, the sultan of Morocco, decided that all free blacks in his realm should be re-enslaved. He legitimized his actions in documents containing words that that echo Ibn Kalduhn: "blacks have natural, good qualities as long as they are in a servile status. Once free, they would return to their natural state of corruption and irreligion. These texts therefore tacitly imply that blacks are natural slaves!" (El Hamel 2002:47). Once again, this sound very much like the complaints about African Americans heard in the American South after emancipation and the use of the notorious Black Codes to control, restrict, and inhibit the movement and behavior of ex-slaves (Walsh & Hemmens 2011).

Is Race a Socially Dangerous Idea?

Conveniently ignoring the contrary 1951 UNESCO statement on race, Yee and his colleagues (1993:1132) inform us that, "Mindful of World War II, UNESCO worked to debunk the idea of race as a biological fact so that it could never again be used to support aggression and genocide. The 1950 UNESCO statement recommended that the term of *ethnic* replace *race*." Following this reasoning to its conclusion, if we could eliminate the term race, we would eliminate a major support for acts of war and genocide. If only things were that simple!

This is a strange kind of logic invoking a form of linguistic determinism and reverse causation. The claim that the race concept was invented to justify foul practices such as slavery and colonialism has morphed into the claim that the concept is the cause of nefarious practices such as aggression and genocide. If there were a shred of evidence that denying the reality of race would prevent hatred, genocide, and

113

war, all persons of conscience would surely rush to deny it, regardless of evidence to the contrary. But as the bloody history of the world attests, it would not. It was *ethnic* cleansing that motivated the slaughters in the former Yugoslavia and in Rwanda at the end of the 20th century, as well as countless other conflicts throughout history. Surely nationalism and religion have been the main culprits in humanity's bloody history. Jonathan Marks arrives at the same conclusion, noting that "If biologically diverse peoples had no biological differences but were marked simply on the basis of language, religion, or behavior, the same problems would still exist" (1996:131). Substituting *ethnic* for *race* does nothing but replace one term with another; it does not shift the reality underlying them, nor do official professional fiats.[2]

Contrary to many constructionists claims, even members of the 1950 committee did not go so far as to say that race does not exist or that the concept had no utility in science; rather, they were arguing for a re-conceptualization of it (Reardon 2005). In this light it is interesting to see what the *rapporteur* of the 1951 UNESCO statement, biologist Leslie Dunn, had to say about his committee's final statement. We saw earlier that Graves (2001) ascribed the contrary statements of the 1950 and 1951 committees to their respective political ideologies, but their disciplinary composition is more salient (how did Graves know the political leanings of members of the committees?). The 1950 committee had only one natural scientist on it, while the 1951 committee had 8 out of 11 (UNESCO 1969). This is doubtless why, rather than denying race, the 1951 committee asserted its dynamic (non-essentialist) nature:

> We were careful to avoid dogmatic definitions of race, since, as products of evolutionary factors, it is a dynamic rather than a static concept. . . . The physical anthropologists and the man in the street both know that races exists; the former, from the scientifically recognizable and measurable congeries of traits which he uses to classify the varieties of man; the latter from the immediate evidence of his senses when he sees an African, a European, and Asiatic and an American Indian together. (Dunn 1969:37)

Perhaps the term *race* has more sinister undertones than *ethnicity*; after all, Hitler used the term. For many social scientists, any sort of biological explanation seems more mysterious, powerful, and threatening than social explanations, because of their erroneous tendency to believe that biology implies fixation. The horrors of Nazism are endlessly invoked as exemplifying the dangers of biological theories, even though similar nightmares have bedeviled humanity throughout its history, none of which waited for Gregor Mendel or Charles Darwin to sanctify them. Nazi theories of racial superiority did not rest on any kind of reputable science, but rather on a quasi-mystical Teutonic nationalism that hypnotized the German people. While the Nazis tapped ancient underpinnings of tribalism and xenophobia to mobilize the German people to their purpose, the mechanisms allowing them the access these traits were social and psychological. The Nazis had control of the media and all social institutions and staged frighteningly magnificent rallies that cemented the cult of the *Füehrer*, fed nationalism, and awaked the monsters of Aryan racial purity and out-group hatred.

History is a sad catalogue of inquisitions, gulags, pogroms, genocides, and wars fought in the name of religious and secular ideologies far removed from any whiff of the demon biology. The communist terror was both longer-lived and quantitatively more heinous than the Nazi terror. The various programs of extermination carried out in the Soviet Union, China, Cambodia, and other so-called workers' paradises were based squarely on a well-articulated theory of human nature that was purely environmental and that explicitly repudiated biology. The Marxist terror did not rest on myths of racial superiority but on myths of egalitarianism. In addition to being scientifically untenable, the *tabula rasa* view is disrespectful of human dignity in that it views us as mere pawns of the environment, waiting to be molded into any shape our cultures might desire, which is precisely why so many have called it a dictator's dream. In light of this tragic history, it is puzzling to see biology pilloried as the bad guy of human rights and human progress by many well-meaning social scientists who have been characterized as being not only oblivious to biology but "militantly and proudly ignorant" (van den Berghe 1990:177).

Race and Racism in the Modern United States

According to Tuch and Martin (1997), most modern commentators maintain that old-fashioned Jim Crow racial attitudes among contemporary white Americans are practically dead. Bobo and Kluegel (1997:93) point out that "Most whites now endorse integration in principle and reject discrimination, preferring instead equal treatment regardless of race. Most whites also deny that blacks are innately inferior to whites." Racial equality in the civil, political, and social realms is now a fact enshrined in the law of the land and accepted by all but the most reactionary racists. Although more African Americans are in poverty than are members of other races, African American economist Glen Loury writes that "we are the most privileged, empowered people of African descent anywhere on the globe" (1995:200), and another African American economist, Walter Williams, writes similarly: "Blacks spend enough money each year to make us, if we were a nation, the 14th richest" (2002:2).

We have a national holiday celebrating Martin Luther King, Jr., the only individual of any race so honored, since even George Washington has been absorbed into a generic President's Day. We see blacks winning mayoral races across the United States in cities where blacks are a minority of the population (Thernstrom & Thernstrom 1997). We have had two successive African American Secretaries of State, two Supreme Court Justices, and countless others in high ranking positions in government, military, business, and education.[3] The election of Barack Hussein Obama "whose name shouts its Third World otherness" (Harris and Davidson 2009:2), to the Presidency speaks volumes about current racial attitudes. Obama garnered more of the white vote than Democratic candidate John Kerry received in the previous presidential election (Caswell 2009). Some have even claimed that Obama won precisely *because* he is black (Ansolabehere & Stewart 2009). Obama's razor-thin résumé and his friendly association with radicals such as Jeremiah Wright, Bill Ayers, Bernadine Dohrn, and Michael Pfleger, all of whom have repeatedly expressed anti-American sentiments, would have sabotaged the campaign of any white candidate.

However, even as they try to kill race, social scientists with a vested interest in keeping racism alive are busy trying to resurrect it by constructing new forms with names like symbolic racism and laissez-faire racism, which are supposedly expressed subconsciously.[4] Racism is under new management, and a research program has arisen under it devoted to demonstrating its continued existence. The expressions of these new racisms consist of the endorsement on questionnaires of traditional values (hard work, personal responsibility, marriage commitment, etc.) and resistance to racial preference policies such as affirmative action (Hughes 1997). In this view, there can be no principled argument against race-based policies without proponents of such arguments being condemned as racists.

There is even said to be a form of racism capable of capturing the most liberal of whites called aversive racism. Whites who feel culturally advantaged, or who feel "a mild discomfort, or fear around blacks," are considered aversive racists (Zuriff 2002:121). Lumping together opponents of racial preference programs, supporters of traditional values, and people who are not comfortable around members of another race assures that just about every white person will be defined as a racist. This is a most satisfying state of affairs for those whose careers rest on finding racism everywhere they look, but there is real damage done to race relations in this country by conflating Jim Crow racism with these new washed-out versions. The effect on blacks upon hearing from the classroom podium (doubtless from an aversive racist) that all whites are racist can be nothing but negative. It is difficult for blacks hearing such messages not to become angry and resentful against whites.

Because it is an article of liberal faith that blacks cannot be racist, all subjects used to discover these diluted racisms are white (Zuriff 2002). If black subjects were to be included in these studies, we would probably find that a majority of them would endorse traditional values, that a significant minority would express opposition to racial preference programs, and that at least some would be uncomfortable around their fellow blacks. We might recall Jesse Jackson's admission that, "thinking robbery," he felt relief at discovering that the footsteps he heard behind him in the night belonged to whites.[5] The same indicators of racism used in these studies to brand whites as racist would thus reveal a majority of blacks to be anti-black racists, an absurdity that reveals the absurdity of this line of research.

Conclusion

The concept of race, if not the English word, long predates the 18th century. From our examination of ancient conceptions, it seems that our 18th-century ancestors bought the images of Sub-Saharan Africans wholesale from ancient cultures. Perhaps these images were used to justify slavery, particularly the images pained by Ibn Khaldun and Aristotle of blacks as natural slaves, but is this historical fact sufficient reason to jettison the concept of race? Religion has been used for centuries to justify the persecution of the Jews, history's perennial victims, but we do not hear loud choruses singing the need to rid ourselves of the concept of religion. If we attempted to dump every idea that has been used to harm others, we would have precious few of them left. It would be very nice indeed to get rid of bigotry, whether religious, racial, or any other form, and surely we can think of more reasoned ways to try to do so than denying what we see every day.

The common mantra "Everyone knows that race is a social construct" echoes in social science corridors all across the country, but as we will see, it doesn't seem to have taken hold beyond the Western world. Even in the United States, Ann Morning (2007) finds that it is a minority view in physical anthropology and, particularly so, in biology. She concludes that the belief in sociology that there is a cross-discipline consensus that race is a social construction "represents a blind faith in constructivism that ignores the signs of a resurgence in biology-based race science" (2007:451). Race is rapidly moving from a phenotypic to a genotypic phenomenon, challenging the oft-heard idea that race is only "skin deep." In light of the exponential growth in the genomic sciences, the sterile moralizing of social constructionists seems hardly adequate to the task of responding to it. But if constructionists cannot match the sheer weight of their opponents' intellectual might head on, they still need to be engaged in nipping at their flanks. Keeping science on the alert for guerilla forays is where gadflies of social constructionism can be of assistance, for troops get fat and complacent without an enemy in the woods.

Endnotes

1. Charles Darwin often castigated Britain and America for their role in the slave trade but also realized that they did more than any other country to end it. He wrote, "It makes one's blood boil, yet heart tremble, to think that Englishmen and our American descendants, with their boastful cry of liberty, have been and are so guilty: but it is a consolation to reflect that we have made a greater sacrifice than ever made by any nation to expiate our sin" (in Richerson & Boyd 2010:565). This was written shortly after Britain freed all colonial slaves at a cost of £20 million, which was 37% of the British government's revenue in 1831, and had fought costly wars to end the slave trade since 1807 (Blackburn 2000). A quarter of a century later, the United States would do the same thing at even greater cost.

2. A shift paralleling the biological sex/cultural gender divide has emerged in many anthropology textbooks in which *race* has been replaced with *biological race* to designate its biological underpinnings, and *ethnicity* used to designate cultural groups. But just as sex and gender are inextricably interwoven (we put down *male* or *female* on questionnaires regardless if they ask for our sex or our gender), so would race and ethnicity be if used in this way, since we think of ethnic groups in terms of nationality rather than in terms of continental ancestry.

3. Far from being excluded from good jobs by racist institutions, thanks to a variety of racial preference programs African Americans are greatly over-represented in government jobs. While constituting just under 13% of the U.S. population, they are 25% of the employees at Treasury and Veteran's Affairs, 31% of the State Department, 37% of Department of Education, 30% of Housing and Urban Development, 42% of the Equal Opportunity

Commission, 55% of the Government Printing Office, 44% and 50%, respectively, of the quasi-government organizations Fannie Mae and Freddie Mac, and an astounding 82% of the employees of the Court Services and Offender Supervision Agency (CSOSA) (Buchanan 2011). CSOSA lists 100% of its first level managers and 67.56% of its senior level managers as black (USOEEC 2009). The U.S. Office of Personnel Management (2006) lists many other federal agencies where blacks are greatly overrepresented, sometimes as large as by a factor of 5. If this is racism (at least if it's anti-black racism), they will have to revise the dictionaries.

4. A number of more conservative black leaders have commented on the role of the self-appointed leadership of the black community in perpetuating a cult of victimhood. In the early 20th century, Booker T. Washington (1972:30) wrote:

    > There is another class of colored people who make a business of keeping the troubles of the Negro race before the public. Having learned that they are able to make a living out of their troubles, they have grown into the settled habit of advertising their wrongs—partly because they want sympathy and partly because it pays. Some of these people do not want the Negro to lose their grievances, because they do not want to lose their jobs.

5. Jesse Jackson is certainly not the only African American to "feel unease" around his own people. Anthropologist and college president Johnette Cole has written that among black women "one of the most painful things I hear is: 'I'm afraid of my own people.'" Criminologist William Oliver writes that, "in response to the prevalence of violence in their communities, many blacks manifest an overt fear of other blacks" (both cited in D'Sousa 1995:267). Can we call these black folks anti-black "aversive racists" because of their unease around their own people?

Chapter 10

The Enlightenment and Scientific Racial Classification

Major Definitions of Race

According to Michael Woodley, there are four major definitions of race in the scientific literature: essentialism, taxonomic, population, and lineage. According to one's outlook, these definitions differ significantly or minimally, but Woodley considers the various views to be "synonymous within the context of the framework of race as a correlation structure of traits" (2010:195). Woodley is saying that if we ignore the nuances, the different methods employed, and edit out all the qualifiers, they all converge on a similar image. That is, they all retain the idea of biologically distinguishable subcategories of *Homo sapiens*. Similarly, Neven Sesardic (2010:144) writes that there are "three grades of racial differentiation that are rooted in biology: genetic, morphological, and psychological." According to both Woodley and Sesardic, then, all biological views on race start by noting that diverse physical, psychological, and behavioral traits among human groups tend to be correlated and then try to explain why. I address morphology and psychological/behavioral traits in this chapter and genetics in the next.

Folk concepts of race have doubtless existed as long as people have noted that others look different from them. "The art of the ancient civilizations of Egypt, Greece, Rome, India, and China, and the Islamic civilization from AD 700 to 1400 shows that these societies classified the various peoples they encountered into broad racial groups. They sorted them based upon the same set of characteristics—skin color, hair form, and head shape—allegedly constructed by Europeans when they invented 'race' to justify colonialism and white supremacy" (Sarich & Meile 2004:30). However, *scientific* subdivision of humans based on shared inherited traits did not exist until the 18th century.

The 18th century marked a change in so many areas on inquiry because it was the century of that broad cultural movement called the Enlightenment. If forced to choose a starting point for the Enlightenment, we could not do better than the late 1680s. In 1687, Newton's *Principia* was published, and in 1689, the supremacy of Parliament over the monarch in England was marked by the enactment of the English Bill of Rights. The Enlightenment was thus a period of liberation marking the beginning of humankind's freedom from the shackles of mind and body it had long endured. Enlightenment thinkers emphasized the use of reason to challenge all previously accepted traditions in religion, philosophy, and government, ushered in an age of secularism and humanitarian reforms, and had an almost religious reverence for science.

With such enthusiasm for science and with the secularization of society, it is not surprising that scientists developed a desire to situate humanity within the sphere of the natural. Perhaps on religious grounds, prior to the Enlightenment the classificatory schemes of biology had been limited to flora and fauna. Humans had theretofore been considered beings temporarily inhabiting physical bodies that they would discard after three score years and ten to meet their destiny in the hereafter, and not at all members of the animal kingdom. As well as freeing scientists from transcendental views of humanity, the Enlightenment fed the natural desire to make sense of the increasingly wide variety of humans being encountered by explorers during this period. It is in the nature of the scientific enterprise that once they are made aware of things that are similar in some ways and different in others, scientists should go about sorting and sifting these things into logically differentiated piles so that they can make sense of them.

According to Hudson (1996:252), the first scientific attempt to divide humans into categories was made by Francoise Bernier, who published a journal article in 1684 in which he distinguished four "species or races of men" (*Especes ou Races d'hommes*). It is generally agreed, however, that Swedish naturalist Carolus Linnaeus (who gave our species its name) is most responsible for our modern idea of race as a classification of humans. Classification of living things is called taxonomy, and Linnaeus was a taxonomist *par excellence*. In his *Systemae Naturae* (published in 1735) Linnaeus presented the world with four subspecies, races, or varieties (he favored the term varieties) of *Homo sapiens*: *Africanus, Americanus, Asiaticus,* and *Europeanus*.

Taxonomy is the art of sorting and naming living things according to some logical scheme. It relies on the scientific data at hand but is still an art and not a science. Thus, different taxonomists may take the same data and arrive at a different number of classifications with different names and determined by different subsets of data. Nevertheless, they all seemed to identify and define subspecies, which in biology is a synonym for race, in ways that depart little from the modern definition given by Patten (2010:67): "[A] subspecies represents a level of biological organization below the species level that has phenotypic properties that are sufficiently distinct (i.e., separable statistically . . .) from other populations."

Linnaeus was in complete agreement with the consensus of the time that a species was an essentialist fixed and unchanging entity; thus, he fits into Woodley's list as both an essentialist and a taxonomist. The principle argument was about the origins of subspecies, which were identified by phenotypic features (observations about physical appearance, behavior, and personality) rather than the statistical methods applied to genetic variation used today. The argument was between the so-called monogenic and polygenic views. The monogenic view held that the races shared one common ancestral source that diverged from the original template due to local environmental factors such as climate, diet, and cultural practices. Monogenists were essentialist about species, but not about race because they allowed for change in subspecies. The polygenic view was essentialist about both and held that the races had separate origins, and some even opined that they were separate species. No one holds such an idea today; using molecular clocks, geneticists estimate that it would take 0.66 million years for hominoids to accumulate enough genetic difference to speciate (Curnoe, Thorne, & Coate 2006).

Linnaeus, and most of his scientific contemporaries, were monogenists, and most were interested in explaining the origins of different human types (how they came to be different from the original type) rather than classifying them. John Greene (1954) provides us with some remarkably prescient speculations on the part of late 18th- to early 19th-century thinkers grasping for pre-Darwinian evolutionary explanations for racial origins. In 1813, William Wells wrote: "Of the accidental varieties of man . . . some of whom would be better fitted to bear the diseases of the country . . ." and "would consequently multiply" (Greene 1954:36). In 1745, Pierre Maupertuis wrote: "[T]he seminal liquor of each kind of animal contains an innumerable multitude of parts appropriate to form their assemblage animals of the same kind" (Greene 1954:35). There is much more in the works of Wells and Maupertuis that is reminiscent of modern concepts such as natural selection, sexual selection, fitness, genetics, and genetic mutation leading to the divergence of different breeding populations from the original common type. There is little evidence of racism in the writings of these early thinkers (of course, determined readers can always find it if they want to) except for the fact that they placed *Europeanus* (or *Caucasians* as they were later labeled) at the top of the hierarchy. This ordering does not imply hatred of other groups. It was for them an honest and accurate appraisal of the relative cultural achievements of the groups. Smug and arrogant? Yes. Hateful and odious? No more than black and gay pride parades imply hatred of whites or heterosexuals, respectively.

The American Anthropological Association (AAA), in its 1998 Statement on Race, makes the point that "Humanity cannot be classified into discrete geographic categories with absolute boundaries." The statement sets up a straw man with its use of the terms discrete and absolute. If races have to be defined as discrete units, then there are no races, since there are no sharp cutting points "with absolute boundaries," even along the species continuum. But who says that they have to be defined this way? No one since the polygenists of the 18th and 19th centuries has subscribed to the demonstrably false assumption that races have to be pure.

Complaints are also often made about the artificiality of racial classification by noting that we have been presented us with as few as three and as many as 63 races. One of the problems of classification is indeed how fine a line to draw between categories. But taxonomical fissioning and fusing occurs in every field of human inquiry without casting doubt on the underlying reality of that which is categorized. Linguists have multiple classificatory schemes that range from a handful of language families to thousands of local dialects. The value of a particular language classification (or any other classification) depends on its purpose.

Morphology and Race

Early taxonomists relied heavily on morphology to classify peoples into races, but how useful is that in science today? Morphology is the form and structure of an organism—how it looks. Michael Woodley (2010) claims that our species possesses high levels of morphological diversity compared with many other species acknowledged to be polytypic (as opposed to monotypic—a single type) with respect to subspecies. One way of distinguishing a subspecies is the so-called 75% rule of thumb. If 75% or more of individuals from a putative subspecies can be distinguished by sight alone from 99% of members of another population of the same species, they constitute a subspecies (Groves 2004). I think it more than possible that almost all adults could classify human individuals correctly according to this criterion more than 75% of the time. Woodley (2010) writes, by way of contrast, that the four recognized subspecies of chimpanzees are much more difficult to differentiate by visual inspection (at least by humans; maybe the chimps would say the same about us—we all look alike). Similarly, Vincent Sarich, a major figure in population genetics, writes, "I am not aware of another mammalian species whose constituent races are as strongly marked [morphologically] than are ours . . . except, of course, domesticated dogs" (Sarich & Meile 2004:170). In other words, Sarich is asserting that naturally selected physical differences among human races are greater than for any other non-domesticated species, and that we must look to artificially selected dog breeds to find a greater degree of intraspecies morphological variation.

Because they deal with skeletal remains to categorize them as to age, sex, or race, physical anthropologists have the greatest claim to expertise with regard to classification based on morphology when all of the features laypersons use to classify race have decomposed. Depending on the number of variables measured, forensic anthropologists can correctly assign race by skull morphology alone between 80% (two variables) and 98% (13 variables) of the time (Konigsberg et al. 2009). This is certainly a very useful application of racial classification for anyone interested in the identity of human remains. Yet there are physical anthropologists who deny that race exists.

In the previous chapter we saw that according to the historian Bernard Lewis, Western academics carry a burden of guilt about slavery and attempt to make it right by claiming that race and racism are Western inventions used to justify it. This apparent guilt trip is illustrated by the degree to which Western anthropologists say they reject the race concept compared with their Eastern European colleagues. Kaszycka, Strkalj, and Strzatko (2009) found that 67% of Western European anthropologists rejected the idea that there are biological races within *Homo sapiens*, whereas 70% of Eastern European anthropologists accept that there are. Echoing Lewis, Kaszycka and his colleagues (2009:51) suggest that Eastern anthropologists are more open to the race concept because their cultures "were not burdened by past colonialism." Results from other surveys of anthropologists have found that between 41% and 69% of American anthropologists deny the existence of human biological races (Lieberman & Kirk 2002), and that 75% of Polish anthropologists accept human races as a biological reality (Kaszycka & Strzalko 2003). Race is also alive and well in China. Examining a large number of Chinese anthropological articles about human variation, Wang, Strkali, and Sun (2003:403) found that "all of the articles used the race concept and none of them questioned its value."

The fact remains that a significant majority of American and Western European physical anthropologists function within a classificatory system they allegedly do not believe in, although this should be put in the context of the official position of the AAA on race and of the professional consequences of voicing a contrary opinion (and not forgetting the guilt that allegedly grips many Western intellectuals). In the politically charged and intolerant context of American academia, the results of polls on race are highly suspect. Anthropologist Henry Harpending's view on this says it all: "A poll about views on race would be like a poll about Marxism in East Germany in 1980. Everyone would lie" (in Sesardic 2010:157). However, if in their minds physical anthropologists define biological race in terms of the AAA's essentialist absolute purity, they can have their cake and eat it too. That is, they can say with a poker face that the philosopher's "really real" race does not exist, and at the same time use the biologist's "merely real" in their work to identify and categorize human remains.

Evolution and Behavior

When we observe some aspect of human anatomy and physiology, we infer that it was selected over alternate designs because it better served some function useful in assisting the proliferation of its owner's genes. We make that inference because there is no other scientifically viable explanation for morphological design. Neither is there any other scientific explanation for the origin of basic behavioral and psychological trait design, although some social scientists would dismiss the idea as genetic determinism. Commenting on such attitudes, Kenrick and Simpson (1997:1) state that "to study any animal species while refusing to consider the evolved adaptive significance of their behavior would be considered pure folly . . . unless the species in question is *Homo sapiens.*" John Alcock (2001:223) makes a similar point: "To say that human behavior and our other attributes cannot be analyzed in evolutionary terms requires the acceptance of a genuinely bizarre position, namely, that we alone among animal species have somehow managed to achieve independence from our evolutionary history."

Behavioral analysis is at the very heart of evolutionary processes because behavior is evoked in response to environmental challenges, and natural selection passes judgment on behavior that has fitness consequences. Selection for adaptive behavioral traits is almost certainly more rapid than for physical traits because organisms play an active part in the selection of their behavior. This is why Plomin and his colleagues have asserted that "the behavioral genomic level of analysis may be the most appropriate level of understanding for evolution because the functioning of the whole organism drives evolution. That is, *behavior is often the cutting edge of natural selection*" (2003:533). After all, it is behavior that creates new variants, and then and only then can natural selection engage in selective retention of the alleles underlying that behavior. When we speak of the evolution of behavior, we are talking about behavioral traits and general propensities to behave in one way rather than in another, not about specific behaviors; nature does not waste precious DNA coding for specifics.

We noted in Chapter 5 how the rate of human genetic change has been about 100 times greater over the last 40,000 years (the approximate time when groups of *Homo sapiens* moved "out of Africa" to become modern Asians and Europeans) than the previous five million years, and that this is due largely to the greater challenges posed by living in ever larger social groups. Living in social groups involves adapting to others in the group, and that is behavior. We saw how the challenges faced by living in larger social groups and the ecological challenges posed in colder northern climates led to robust increases in cranial capacity. Thus, not only does behavior in various contexts lead to selection for behavioral and psychological propensities, behavior also effects selection for morphological feature. The discovery of gene-brain-culture co-evolution is likely to profoundly influence the study of both human origins and human behavior (Walsh & Bolen 2012). Because the three traditional anthropological races—Asians, blacks, whites—evolved under different climactic, ecological, and social conditions, evolutionary logic demands that different sets of overlapping physical, psychological, and behavioral traits will have evolved that reflect those conditions. Only those who disavow evolutionary logic and/or believe that humans have somehow escaped the effects of natural selection can deny this in good conscience.

Behavioral and Psychological Traits and Race

Behavioral and personality traits were also used by early taxonomists to define race. The usual procedure was to note a person's race and then to correlate it with his or her traits and behaviors. This is where accusations of racism are most likely to be evoked. Racism, in part, involves *falsely* attaching cognitive, temperamental, and moral characteristics to phenotypic features: for instance, observing that such and such a person is Asian, black, or white and then making certain false assumptions based on that. If those assumptions are supported by multiple lines of robust evidence, however, is it still racist to note them, regardless of how unpalatable they may be? Is it racist even to ask the question? Even if the evidence for meaningful racial differences across many life domains is overwhelming, some may feel it better for one reason or another to hide it than to reveal it. These folks share an ancient tradition with those well-intentioned souls who pilloried iconoclasts such as Copernicus and Darwin for upsetting comforting images about ourselves.

Suppose we obtain a wide variety of heritable physical, psychological, and behavioral measures such as those presented in Table 1 from many thousands of individuals of African, Asian, and European ancestry. We assume that these measures constitute discriminating variables measuring characteristics on which the groups are hypothesized to differ. We then perform a discriminant analysis, the success of which depends on the mean values of the predictor variables being significantly different among the groups. If these measures vary randomly across the groups, then the analysis will fail, and we conclude that the groupings (races) are not scientifically meaningful. If the measures vary systematically, it will succeed, and we can conclude that the groupings are scientifically meaningful and that we can make predictions based on them. How many of those who believe that race does not exist (or that it is not a useful concept) would bet on the failure of the analysis? There would be much overlap in the group distributions on all variables, and thus that the groups would not be pure in any sense. However, if complete statistical independence is the criterion for the usefulness of any taxonomical concept, then we will have to trash all such concepts and find another line of work. Whether we acknowledge it or not, the big three races can be distinguished from one another on a variety of measures that cluster together in non-random ways.

The Canadian psychologist J. Philippe Rushton (1995) has done such an analysis. Rushton positions Asians, blacks, and whites on a large number of traits based on many hundreds of studies from around the world. Rushton finds that these rankings cohere with prediction made by biology's life history theory. Life history theory, sometimes also known as r/K theory, or differential K theory when applied to humans, is essentially a theory about reproductive strategies of species, sub-species, and breeding populations based on ecological conditions. These strategies vary on a continuum from extreme mating effort (e.g., oysters) accompanied by numerous offspring and no parental care to intensive parenting effort (e.g., humans) with few offspring. Mating effort is considered a fast life history, or an *r* strategy, and parenting effort is considered a slow life history, or *K* strategy. These strategies evolve in response to ecological pressures of the kind we discussed in Chapter 6 (the Mundurucu and Mundugumor versus the Arapesh and !Kung). All humans are highly K selected, but there is variation within the K strategy, with the most obvious being the male-female difference.

Arguments about where human mating populations may fall on the r/K continuum are anything but polite. The controversy sur-rounding Rushton is largely based on his assertion that racial groups aggregate at different points along the intra-species continuum, with Asians more K–selected than whites, who are more K–selected than blacks, and that the strategies followed by the different groups covary, with many heritable traits helpful in the maintenance of the utilized strategy. If this is so, average differences on r or K traits should fall into predictable patterns. As Table 1 shows, Rushton subsumes many different traits under five higher-order concepts: intelligence, maturation rate, personality/temperament, social organization, and reproductive effort. Included are traits in which socialization is heavily involved (achievement, sexuality, and social organization) and traits (morphology, speed of maturation, gamete production) in which social variables are involved minimally, if at all. He then examines the literature on racial differences relating to these variables. If racial differences are random with respect to indicators of mating versus parenting effort, most research results will be null, and the remainder will be about equally split between negative and positive results. Reviews of the literature have shown that this has not been the case across the hundreds of studies that have examined these variables in a variety of contexts and for a variety of research purposes. There have been a few null results, but the vast majority have been positive (Rushton 2010).

Rushton notes that his theory is not a biological theory, but rather a "mixed evolutionary/environmental" one that "fits the data better than any currently available purely genetic or purely environmental alternative" (1991:126). Such a statement would be taken as an obvious truism and hardly worth mentioning if the animal in question was any other than human. It is also important to note that there is considerable overlap of these *average* differences, but the differences are demonstrably there. Working under the rubric of gene-based evolutionary theory, Rushton has woven a network of meaning-joining diverse phenomena and their correlates. Theories able to incorporate phenomena not previously thought to be related into a coherent explanatory scheme are generally welcomed and much admired in science, but Rushton's work has been demonized as racist and viciously attacked by those who would shackle the mind.[1]

Some of Rushton's critics admit that it is difficult to criticize his work on scientific grounds. For instance, Daniel Freedman states that the assemblage of data on race collected by Rushton is "not readily dismissible, although many have tried" (1997:61). He goes on to state that the major problem with Rushton's work is not with his data, "but with the emotionally distant nature of his scientific presentations" (1997:61). Thus, Rushton is merely "emotionally distant" (Romanticism redux) rather than racist. Another writes that "[It] would take a variety of environmental factors to explain all of the racial differences [that Rushton's theory] parsimoniously accounts for" (Lynn 1989:5). And yet another: "All in all, I find the pattern that Rushton presents interesting and worth pursuing" (Mealey 1990:387).

Table 1
Average Differences Among Blacks, Whites, and Orientals

Variable	Blacks	Whites	Orientals
Brain size			
Cranial capacity (cm^3)	1,267	1,347	1,364
Cortical neurons (billions)	13,185	13,665	13,767
Intelligence			
IQ test scores	85	100	106
Decision times	Slower	Intermediate	Faster
Cultural achievements	Lower	Higher	Higher
Reproduction			
Two-egg twinning (per 1,00 births)	16	8	4
Hormone levels	Higher	Intermediate	Lower
Secondary sex characteristics	Larger	Intermediate	Smaller
Intercourse frequencies	Higher	Intermediate	Lower
Permissive attitudes	Higher	Intermediate	Lower
Sexually transmitted diseases	Higher	Intermediate	Lower
Personality			
Aggressiveness	Higher	Intermediate	Lower
Cautiousness	Lower	Intermediate	Higher
Dominance	Higher	Intermediate	Lower
Impulsivity	Higher	Intermediate	Lower
Self-concept	Higher	Intermediate	Lower
Sociability	Higher	Intermediate	Lower
Maturation Rate			
Gestation time	Shorter	Longer	Longer
Skeletal development	Earlier	Intermediate	Later
Motor Development	Earlier	Intermediate	Later
Dental Development	Earlier	Intermediate	Later
Age at first intercourse	Earlier	Intermediate	Later
Age of first pregnancy	Earlier	Intermediate	Later
Life span	Shortest	Intermediate	Longer
Social Organization			
Marital stability	Lower	Intermediate	Higher
Law abidingness	Lower	Intermediate	Higher
Mental health	Lower	Intermediate	Higher

Source: J. P. Rushton (2000). Race, Evolution, and Behavior. 2nd Abridged edition. With permission.

I initiated a thread in the Human Behavior and Evolution Society's (HBES) e-mail discussion list in 2003, asking if anyone was aware of any alternatives to Rushton's theory that could explain the consistent racial patterning he identifies. The thread developed a lively discussion, but no one knew of an alternative. There was considerable agreement that if opponents really believed their arguments, they would be in the forefront collecting comparative racial data to support them. Rather than do this, however, their main tactics appear either to come up with ad hoc explanations specific to time or place, to avoid collecting such data at all, and to intimidate others into doing the same. Rushton thus wins by default, because his is the only show in town. This does not mean that it is the final word in understanding racial variation from an evolutionary perspective, or that one cannot find flaws in it, which is to be expected in any theory, especially in theories of such broad scope.

Whatever one thinks of the merits of Rushton's work, it remains true that none of his critics have supplied aggregate data indicating anything other than the racial gradient he identifies. It is almost impossible to imagine what strictly environmental factors could account for the systematic alignment of the three racial groups on such a wide variety of traits consistently documented across cultures. This is the crux of the matter, and whether or not Rushton has a racist agenda is irrelevant to the substance of his scientific work. His work is exemplary; if it were not, he would not have close to 300 peer-reviewed journal articles listed in his online vita, nor would he have been elected a fellow of the American Association for the Advancement of Science or the Guggenheim Memorial Foundation. The vast majority of empirical studies testing life history theory among humans have been supportive (e.g., Figueredo et al. 2006, 2007; Griskevicius et al. 2011; Jonason, Koenig, & Tost 2010).

Rushton's work does not consider the sensitivities of those who take offense at it, but science, the journals that published it, and the universities where it is taught are not in the business of assuring all people that their sensibilities will not be disturbed. If Rushton's facts are wrong or his theory illogical, that is his problem; if his theory offends you, that is your problem. The pursuit of knowledge about human nature and about our place in the universe has never been confined to polite chatter. Those opposed to research on racial (or gender) differences should be reminded that the worth, dignity, and essential equality of human beings is a moral axiom that is in no way contingent on findings of average differences among groups. The extent to which we afford worth, dignity, and respect to individuals should depend on the content of their characters, as Martin Luther King reminded us, not on the groups to which they belong.

Anti-White Racism and White Studies

Given the accusations of racism thrown at Rushton and others like him, it is instructive to revisit and expand on the theme of reverse ethnocentrism touched upon in the last chapter. There are examples of anti-white racism that appeared in a now defunct journal actually devoted to race hatred called *Race Traitor*, edited by historian Noel Ignatiev. The journal's mission statement says it all:

> RACE TRAITOR aims to serve as an intellectual center for those seeking to abolish the white race. It will encourage dissent from the conformity that maintains it and popularize examples of defection from its ranks, analyze the forces that hold it together and those that promise to tear it apart. Part of its task will be to promote debate among abolitionists. When possible, it will support practical measures, guided by the principle. Treason to whiteness is loyalty to humanity. (http://racetraitor.org/)

Although Ignatiev and his fellow "traitors" did not literally mean that they desire the mass extermination of whites (although one critic called it "worryingly reminiscent of Nazi propaganda" [Cole 2009:33]), the journal contains article after article vilifying the white race and blaming whites for every evil in the world. One essay by history professor Grace Hale concludes by asking rhetorically if America would be America without white people, to which Hale replies, "No. It would be something better, the fulfillment of what we postpone by calling it a dream" (cited in Kolchin 2002:168). I do not know the races or nationalities of the authors (Hale is white) who contributed to this journal, but I assume that the majority are white Americans. These folks go beyond reverse ethnocentricity to reveal a loathing of their race and nationality and, because race and nationality are at the core of one's identity, a loathing of themselves. Ignatiev, for instance, distances himself from his heritage by saying he is not a Jew but "has a Jewish background" (Robertson 1997:12).

Ignatiev is considered to be the father of so-called whiteness studies in academia. Whiteness studies is part of the deconstruction of race movement and is characterized by David Horowitz as smacking of self-hatred. He points out that "Black studies celebrates blackness, Chicano studies celebrates Chicanos, women's studies celebrates women, and white studies attacks white people as evil" (in Hartman 2004:23). This opinion is shared by Barbara Kay (2006:2), who writes, "That elite universities where WS [white studies] is taught, such as

Princeton and UCLA, tolerate their students' being indoctrinated with Ignatiev and company's pernicious race theories points to a new low in moral vacuity and civilizational self-loathing."[1] We saw in Chapter 2 how Bruce Charlton considered the clever silly state to be a tragic one, but here is clever silliness combined with needless guilt that has reach self-loathing proportions. Or perhaps it is not self-loathing but rather a narcissistic desire to feel superior to mere mortals, a symptom demonstrated most publically by the not-so-clever silly Michael Moore with his *Stupid White Men* (2001), a book that received rave reviews from leftist types. Such silliness is not only tragic for those mired in it, but it contributes to poor race relations to the extent that non-whites buy into it and to which whites react to it by pointing out negative aspects of other races in self-defense. Surely Ignatiev and company are aware that people from all corners of the earth risk all to come to "evil white countries," because it is in those countries where rights and liberties are most respected, and where one's material needs are best met.

As far as I am aware, no one has been attacked, censured, fired, or otherwise harassed for contributing to *Race Traitor*, and to my knowledge there have been no attempts to censor it. As previously noted, it is now defunct, but calls have been made to resurrect it (Preston & Chadderton 2012). One can only imagine the fate of a journal devoted to abolishing the black race (if only metaphorically), or of the authors who contribute to it. Yet serious scholars of race such as Rushton are continually attacked for their empirical work published in top-tier scientific journals (not bottom-tier rags with an obvious agenda like *Race Traitor*), and he has even threatened with criminal charges for doing so by Canadian authorities for "inciting racial hatred," the evidence being Rushton's peer-reviewed scientific work! (Kohn 1999).

Conclusion

For better or for worse, beginning with the Enlightenment scientists began to situate human beings in the realm of the natural. Part of this process was to classify human groups into kinds, as is done with flora and fauna. This was done initially by looking at physical characteristics and behavior. We have seen that an essentialist position on race has always been the minority view among scientists, who have always declared that people differentiate according to the environmental challenges they meet in their long evolutionary history. Those with an unconstrained vision have never liked the idea of meaningful group differences in behavior, personality, and cognition, even though evolutionary biologists would be more than just surprised if there were none, given the logic of natural selection.

From the works of Copernicus, to Darwin, to Rushton, people have been hurt and even outraged when their intellectual comfort zones are invaded, but that is the price we pay for furthering our understanding of ourselves. What a dark hole of ignorance we would find ourselves in if all the politically incorrect bad guys in science had been intimidated into silence. If the attitude prevailed that all pronouncements about humanity must be exquisitely sensitive to everyone's feelings, never offensive or hurtful to anyone, we would still hold the belief that we are at the center of the universe and that we are more like angels created in God's image than animals created by nature. While such a belief is comforting and cozy, it does not match what the cold stare of science reveals. Furthermore, charges of racism aimed at Rushton based on his scientific work in which not a single word of racist rhetoric is found must be balanced against the real racial (self) hatred appearing in *Race Traitor* and various other venues to which academics contribute.

Endnote

1. The United States is not alone in suffering the inanities presented in whiteness studies. Michael West (2005:5) recounts how schools in the U.K. do not emphasize the rich traditions of Western culture; rather, they celebrate non-indigenous cultures, a celebration that is often accompanied by the:

    > simultaneously belittlement of British culture. History in schools concentrates not on the UK's role as a pioneer of parliamentary democracy, how it was one of the first countries to abolish slavery, how it has been a place of refuge for minorities fleeing persecution. . . . Rather, we are told to hold our heads in shame at our nation's abominable record of colonialism and oppression.

    He goes on to echo the self-loathing theme of the American, Horowitz, and the Canadian, Kay, to say, "The failure of the socialist projects of the twentieth century led many on the left to give up any hope. . . . [W]e have become self-hating, deeming Western man as an agent merely of war, racism, slavery . . ." (2005:6).

Chapter 11

Race and Molecular Genetics

Essentialism Revisited

Now that we are getting to the crux of the race matter in terms of molecular genetics, it is a good time to revisit essentialism, the social constructionist's weapon of mass destruction. Constructionists use it to demolish the opposition without fear of contradiction, because by claiming that race realists hold essentialist positions, they poison the well with definitions of race that their proponents cannot recognize. For instance, philosopher Naomi Zack claims that believers in the reality of race "to this day . . . assume the following (1) races are made up of individuals sharing the same essence; (2) each race is sharply discontinuous from all others" (2002:63). I have never heard a natural scientist claim that individuals in any racial group share an "essence" or that racial boundaries are "sharply discontinuous," and neither, I'll wager, has Zack. But it is sad to say that whenever the essentialism charge is made, the folks singing from the same hymnal will give it a hearty amen regardless of how often it has been refuted. These people are "ignoring the dog that is loudly barking here and seeking attention: a biologically informed but *non-essentialist* concept of race" (Sesardic 2010:146).

A team of British sociologists were also deaf to the dog when they insisted that sociologists should not use the race concept because "it has been widely held that genetics cannot substantiate the idea of discrete 'races' and the concept has no scientific validity. This has long been the view of British social science (although not necessarily medical and biological sciences) and is virtually undisputed today" (in Skinner 2007:937). Skinner notes the obvious absurdity in stating that sociologists should not use a concept because it does not exist while admitting that it is used by more advanced sciences. Medical and biological scientists realize that there is no such thing as a race

137

gene, present in all members of one population and absent in all members of other populations. If this is what those who deny the reality of the race concept mean by race, they are laying siege to a castle that surrendered two centuries ago, for it is universally acknowledged in science that *Homo sapiens* constitutes a single taxonomic biological unit—the human race.

All this talk about essentialism, purity, and absolutism—from the high and mighty AAA's dogmatic statement down to the most recently minted PhD constructionist's classroom cant—is a semantic cheat that forces a Manichean choice between either "pure race" or "no race." This argument destroys a straw man, for all scientists acknowledge that there is no human gene pool unadulterated by genes from another pool. It is patently obvious that there are no pure races; if there were, we would call them species. There are no pure languages either. All languages have evolved from a few proto-languages, and all have intermingled promiscuously, yet this does not detract from the concept of language, or from naming the various languages according to their regions of origin. The region of origin of peoples—their ancestry or lineage—is how race is defined, not by its genetic purity. The race concept has real denotative and connotative properties that have proven useful in the medical, forensic, and behavioral sciences, and those who believe this know that all human alleles are present in all human groups. However, the frequency and functional properties of the alleles (variants of the same gene; one from each parent) differ widely among these groups.

Modern science views race in terms of stable differences among different populations of individuals whose ancestors evolved under different environmental conditions. This is consistent with Templeton (1998:646): "A more modern definition of race is that of a distinct evolutionary lineage within a species." Ernst Mayr, the doyen of population genetics, defines race similarly: "A human race consists of the descendants of a once-isolated geographical population primarily adapted for the environmental conditions of their original country" (2002:91). He also states that those who deny that there are no human races "are obviously ignorant of modern biology" (2002:89). This is a tad harsh, for there are biologists who continue to deny the existence of race, even as they investigate the genetics of ethnic groups or populations as surrogates.

It is thus plain that biologists are fully aware that racial boundaries are ambiguous and shifting: relative and dynamic rather than absolute and static. It may be more fruitful sometimes to think in terms of clines (a graded series of morphological changes along lines of geographical transition, much like weather isobars) rather than races; it all depends on the degree of lumping or splitting required by the research issue. Cummings (2000:457) jumps on the cline concept to discount race: "If humans are to be divided into racial groups, large-scale genetic differences should occur along sharp boundaries." Cummings has dredged up essentialist images again as a base from which to wriggle away from the race concept, but he should know (he is a biologist) that we do not observe "large-scale genetic differences along sharp boundaries" even between species. Red and ultraviolet are discernible colors on the color spectrum regardless of the lack of sharp boundaries between the intermediate colors, just as Norwegians and Zulus are discernibly different despite the relative smoothness of the clinal spectrum separating them. Likewise, sociologists sometimes conceptualize social class as a continuous variable and at other times they impose sharp boundaries (e.g., low/middle/high). Is social class a useless concept because of its cline-like tendency to merge smoothly from case to case across the distribution, or because its discrete categories are determined by researchers according to their research purposes, and are definitely not pure?

How Does One Prove that Something Does Not Exist?

Before examining the genetic data, I must address the constructionists' argument that race does not exist more deeply, since all we have encountered so far are bald statements to that effect. Sociologist Bob Carter (200:85) emphatically states that race is "demonstrably false" and then continues to say that its falsity can only be demonstrated "through scientific discourse." One would naturally expect such a discourse to follow, but it does not. We are treated instead to a brew of incestuous quotations and dogmatic pronunciations from Carter's fellow travelers. But how could Carter or anyone else prove that something doesn't exist? After all, it is a principle of folk logic that one cannot prove a negative. He could talk about the absence of sufficient evidence showing that something *does* exist, but how insufficient must the evidence be to claim that it *does not*? The quintessential example of this has been God's existence, but God is in Plato's supernatural realm and beyond the methods of science. Race is in Aristotle's natural realm of the senses, and thus its existence or non-existence can readily be addressed by science.

Because the entity that supposedly does not exist is biological race, we have to turn to biological entities to address the question. Now, to show that something does not exist is actually easier in many cases than to show that it does. Think of a jury trial; the defendant needs only one person to hold the negative position (the defendant is not guilty) to result in an acquittal through a hung jury. The burden of proof is always on those who assert the affirmative (the man is guilty, God exists, race is a reality), not on those who assert the negative. The beauty of this is that every time one proves a positive, one also disproves its negative. When the jury brings in a guilty verdict, the prosecution has both proved the affirmative and disproved the negative; in other words, the assumption of innocence has been rejected beyond a reasonable doubt. The jury hopefully based its decision based on a thorough examination of the facts. This is an inductive process that is not based on preconceived notions, and inductive conclusions strongly suggest truth but do not guarantee it.

To prove that race exists beyond a reasonable doubt, geneticists would have to collect DNA data from a large number of individuals from around the world. They would then test this evidence for certain features that differ among people who come, or whose ancestors came, from different geographical regions of the world. The null hypothesis is the scientist's version of the prosecutor's assumption of innocence. It is the null hypothesis (in this case: "race does not exist") that is tested as a cautionary measure just as the assumption of innocence is used in the criminal justice system, because scientists do not want to make claims that cannot be substantiated any more than society wants to convict innocent people. Of course, neither the scientist nor the prosecutor believe their assumptions (the innocence of the defendant, or the null hypothesis is true), but set them up to require stringent evidence to reject them. If these genetic features (genetic polymorphisms) differ significantly and cluster together according to ancestry, then we reject the null hypothesis that races do not exist. Thus, scientists test the null hypothesis that something does not exist in order to show that it does.

If the null hypothesis cannot be rejected, scientists have "proved" that the negative is more likely than the affirmative. The null hypothesis that race does not exist would be accepted if the genetic measures clustered randomly rather than systematically. Race deniers thus have a perfect opportunity to prove their negative claim provided by the usual methods of science. As the evidence will show below, whenever genetic data have been subjected to analysis, it invariably and robustly clusters into grouping that the Enlightenment taxonomists would readily recognize. Of course, scientists never claim they have proved something unequivocally and absolutely; they only conclude that the evidence for such and such is robust, and true beyond a reasonable doubt.

Another way of explaining the nonexistence of something is to show that it has no physical referents. Phlogiston, for instance, was a hypothetical property said to be in all corrosive and combustive materials that was released when such materials corroded or combusted and were absorbed by the air. There were a number of good logical (but not empirical) reasons supporting the existence of phlogiston, and almost all chemists in the 18th century subscribed to the theory. We now know that corrosion and combustion are processes by which an object combines with the oxygen in the air. Corrosion and combustion thus take from the air rather than give to it. Oxygen is the tangible referent that supports the oxidation theory; there is no tangible referent in chemistry that could be attached to phlogiston that could make it real, therefore it does not exist. Applying this criterion for reality to existence/non-existence of race, we have seen evidence in previous chapters and will see more in this that there are real physical referents to race. If there were none, we could indeed file the race concept in the historical dustbin with phlogiston.

No one is claiming that race is a simple brute fact, however. Brute facts are what we bump into when the chain of scientific regress goes no further, and we are led to say that a fact is just what it is: why does gravity attract rather than repel? It just does. Brute facts can only be described, not explained. Race can be explained by reference to other, deeper, facts, such as natural selection for genes underlying features that were adaptive in different ecological and cultural environments. Race is fluid in that it has been shaped by forces other than biology. Geography, history, and social and cultural forces have driven both the content of the underlying genomes of different racial groups and the content of cultural referents about those groups.

Race and Genetic Distance

The most ambitious work concerning population categorization prior to the completion of the Human Genome Project was Cavalli-Sforza, Menozzi, and Piazza's 1,000-page book *The History and Geography of Human Genes* (hereafter, HGHG) (1994). Their geographic maps of human gene frequencies make it clear that all populations overlap clinally, which clearly shows that genetic differentiation does not place the peoples of the earth into neat discrete piles. They are messy piles, but discernible piles nevertheless. Cavalli-Sforza, Menozzi, and Piazza's work has been widely cited as providing proof that races do not exist, which is doubtless the result of their own assertion that "The classification into races has proved to be a futile exercise" (1994:19). However, the remainder of the book belies their denial, although they never label the groups they identify as races, instead they prefer the term *populations*. Oddly though, the populations they identify based on gene frequencies are referred to in the book by their traditional anthropological racial classifications of Caucasoid, Mongoloid, and Negroid.

This contradiction did not escape critics, who observed that HGHG's message is anything but that of "race does not exist," and that their work "can be summarized in widely publicized color-coded maps in which Africans are yellow, Australians red, [Mongoloids blue], and Caucasoids green" (Harry & Marks 1999:304). Harry and Marks imply that HGHG's "no races" statement is a genuflection to political correctness in an attempt to gain acceptance of Cavalli-Sforza's Human Genetic Diversity Program, which a number of left-leaning scientists oppose as genetic imperialism. Edward Miller (1994) also comments on the apparent contradiction (stating that race does not exist and then proceeding to offer extensive evidence that it does). Cavalli-Sforza (2000:49) makes it clear, however, that it is racial *purity* that does not exist, which we already knew, not race as defined by common ancestry.

HGHG contains genetic data (120 markers from 49 loci) obtained from 42 populations from around the world. The book tells us a lot about the worldwide distribution of a variety of genes, none of which include genes of interest to behavioral scientists, such as those underlying various, cognitive, temperamental, and behavioral traits. Genes underlying these traits are subject to natural selection and are not useful in assessing population genealogies because their frequencies change under selection pressures. The selection-neutral genetic markers used in the HGHG study are passed down across the generations relatively unaltered except for drift and minor mutations, and are thus useful for assessing group genealogies based on genetic distance. Genetic distance measures are quantifications of the relative isolation of breeding populations from one another—the greater the genetic distance, the greater the isolation. The text presents a series of genetic analyses using dendograms and principal component analyses, the results of which cohere well with traditional racial classifications, with African groups clustered in the lower right quadrant, Europeans in the upper right quadrant, and the remaining two quadrants containing Asian and other populations (1994:82).

HGHG also presents matrices of genetic distance for each of 42 populations measured by the fixation index (F_{st}) method.[1] F_{st} is a statistic for analyzing the level of genetic divergence among subpopulations. African Bantus and native Australians show the largest F_{st} value (3272, meaning that 32.72% of their neutral geneses are different), despite having similar physical features resulting from adaptations to similar physical environments, which is why morphology is not always a good indicator of race. The genetic difference between the Bantus and the English is 2288 (22.88%), and between the English and the Danes it is 21 (0.21%). The English are thus 109 times more genetically similar to the Danes than they are to the Bantus. According to fairly well-accepted quantitative guidelines, F_{st} values of .05 to 15 are considered moderate, .15 to .25 large, and $F_{st} > .25$ very large (Nassiry, Javanmard, & Tohidi 2009). The HGHG data support both a clinal and a racial interpretation, depending on whether the research issue requires gross lumping into a few relatively discrete trees, or fine splitting into a clinal forest.

Lewontin Proves a Negative—or Does He?

As well as being used to assert differences among human groups, gene counting has been used to affirm human similarity across populations (it is of course obvious that every human is both similar to every other human and different from all other humans). Arguments against the existence of race almost invariably invoke the one of many iterations of the AAA's official *Statement on Race* and echo the statement that human racial groupings "differ from one another only in about 6% of their genes," and that this small percentage implies the non-existence of race. In effect, this means that there is more genetic variation within races than between them and that the between-groups variation is not large enough to support the race concept. As Klein and Takahata (2002:389) point out, however, such a difference can be enormously significant: "Sewell Wright [the originator of the F_{st} statistic], who can hardly be taken for a dilettante in the question of population genetics, has stated emphatically that if differences of this magnitude were observed in any other species, the groups they distinguish would be called subspecies."

The various "less-than-6%-of-their-genes"-type statements are traced to a study by Richard Lewontin (1972), probably be the most quoted but least understood research in social science relevant to race. This supposed requiem for race was challenged by Jeffry Mitton in 1977, but his work was lost in the euphoria evoked by Lewontin's work. Mitton's (1977) point that multiple genetic loci are superior to a single locus for the purpose of classification was perhaps too obvious for some to grasp. It was not challenged again to my knowledge until statistical geneticist A.W.F. Edwards, who developed many of the techniques on which population genetics depends, did so in 2003. Edwards argued that Lewontin performed his analysis to attack racial classification, which he detested for social and not scientific reasons, and called his work a "superficial analysis" (2003:799).[2] He showed that while Lewontin was correct when examining the frequency of a single genetic locus between individuals (which is expected a priori on probability grounds), the probability of group misclassification rapidly approaches zero when as few as 20 loci are examined. Hacking (2006:85-86) notes that: "Edwards' analysis is, for anyone with a modest statistical training, rather direct and self-evident," and states that Edwards' work "is now widely judged as correct." Thus the negative was not proven after all, although it remains true that there is more genetic variation within races than between them.

Skinner (2007:934) informs us that his fellow sociologists are keen to say that the AAA statements are about "what science tells us about race," but they neglect to add that the science involved in the development of these statements was predominantly *social* science. Social scientists no more possess the training, methods, or means to make credible statements about the existence of race than they had to make authoritative statement about the existence of atoms during the period when physicists and chemists argued about the reality of that concept. After repeating the 6% mantra, the various AAA statements catalog the sad history of racism from colonialism to the Holocaust, all of which is interesting, but it is history, not science. It addresses racism, not race, and then conflates the two terms. There is no scientific justification for racism, but there is for race. It is all too common to conflate the terms and to tar anyone who studies *race* with the *racism* brush and to employ a variety of malicious tactics to try to silence them.

Postgenomic Clustering Studies

The conclusion that races differ by 6% of their genes may be confusing without some background in genetics to place it in context. If humans have 25,000 genes, this does not mean that the races differ by 1,500 genes. Actually, the races do not differ at all in the genes they possess, but rather in the distribution of variants of these genes. Variants of the same genes (alleles) change the way they function and are regulated. By this I mean such things as the quantity of a gene product, how long a gene is turned on, and particularly how they function in the brain to regulate neurotransmitter functioning (the quality and number of receptors, how quickly a neurotransmitter is transported and degraded after it has completed its task, and so on).

To see what is meant by this, suppose we lump all genes from all primate species together such that the pool contains 100% of the genetic variation in the primate order. Regardless of what two primate species we compare, the genetic variation within each is *necessarily* greater than the genetic variation between them. We share 97% of our genes with gorillas, as do chimpanzees, with which we share about 99%, which make us genetically more like chimps than chimps are like gorillas (this approximately1% difference between us and the chimps has not led anyone to conclude that species don't exist). The differences between these species are obvious, but in order to understand them, we can ignore what we share and concentrate on what we do not. Every animal shares the majority of their genes with every other animal, because they share the same evolutionary goals of survival and reproduction and must perform the same tasks that make those goals possible; thus, all require similar DNA.

Although protein-coding and non-coding regions of the chimp and human genomes contribute to functional differences between them, many differences "may have been driven by changes in transcriptional regulation rather than protein function per se" (Babbitt et al. 2010:67). Coding regions and regulatory regions, along with pleiotropic factors (the multiple effects a single gene can have via its interaction with other genes), are all responsible for differences among primate species and among human groups, but many differences in morphology, physiology, and behavior in species, subspecies, and individual organisms are traced to changes in gene regulation (switching them on and off). The same genes code for the same proteins that build necks in geckos and giraffes but are turned on longer in the latter, and both humans and chimps have brain and hair, but humans have more brains and chimps have more hair. Think of gene variants as the same piece of music played by vastly different ensembles ranging from a teenage garage rock band to the London Symphony Orchestra. Nature is parsimonious in that it preserves genes underlying biological processes common to all animals because they work. Natural selection does not create an entirely new genome when species branch off from the parental line any more than authors create new words when they write different books.

As indicated above, all *Homo sapiens* share 100% of their genes but differ on their alleles. These alleles come in a variety of forms called genetic polymorphisms, the two major ones being single nucleotide polymorphisms (SNPs) and micro- and mini-satellites (referred to collectively as variable number of tandem repeats—VNTRs). These minute genetic differences may make an enormous difference at the phenotypic level. For instance, a difference in just one nucleotide is all that differentiates one allele from another in an SNP. The Val158Met SNP of the enzyme catechol-O-methyltransferase (COMT) that degrades a variety of neurotransmitters has the tri-nucleotide sequence *ATG* that produces methionine and a *GTG* sequence that produces valine. In this case we say that there are two alleles for COMT: A and G. This single nucleotide difference is not trivial. The valine variant degrades dopamine (a major chemical reinforcer of pleasure-seeking behavior) at about four times the rate of the methionine variant. This rapid clearance puts val-val homozygotes at risk for addictive pleasure seeking. That is, val-vals get the same dopamine boost that met-met homozygotes get from things that release it, such as food, sex, alcohol, gambling, and drugs, but the dopamine does not linger in their synapses as long, thus driving the often compulsive search for more (Beaver 2009).

VTNRs differ from one another in the length of contiguous nucleotide bases that are repeated a different number of times. The more times the sequence of nucleotides is repeated, the longer the allele. An example of an important VNTR is the 7-repeat allele of the dopamine receptor DRD4 gene. The number of repeats this gene has determines the receptor's sensitivity to dopamine, with shorter (e.g., the 2-repeat) repeats being very sensitive and longer repeats (e.g., the 7-repeat) being much less so. This polymorphism has been associated with many phenotypic traits linked to antisocial behavior such as ADHD and impulsivity (Canli 2006). Thus, it is in allelic variants of the same genes such as SNPs and VNTRs that individual and group genetic differences are found, and these miniscule differences at the molecular level can result in extremely large differences at the level of the organism.

Modern geneticists estimate that any two human beings chosen at random differ only on about 0.01% of their genetic material, although some claim it to be 0.05%. While this 0.01% seems trivial, the human genome contains about 3 billion nucleotide base pairs (the nucleotides TA, CG, AT, etc. are base pairs), so that one-tenth percent represents about 3 million base pair differences. Many of these differences probably have no effect on phenotypic differences, but Bonham, Warshauer-Baker, and Collins (2005:12) state that "about 200,000 common variants are responsible for the genetic components of the differences in health, behavior, and other human traits." And according to Plomin (2003:195), these base pairs are "highly variable between different regions of the genome and in different ethnic groups." What has changed since Lewontin's 1972 study is the realization that however modest the interracial genetic variance may be, it nonetheless represents millions of possible genetic differences and combinations.

The degree of variation in genes affecting quantitative traits (the kinds of traits social/behavioral scientists are interested in such as aggression, conscientiousness, IQ, and empathy) that depend on the cumulative action of many genes often exceed those affecting qualitative traits such as the markers used in HGHG because they are subject to ongoing natural selection (Bamshad et al. 2004). Thus, we would expect them to vary considerably between populations that have experienced different ecological and cultural environments on different continents over evolutionary time. There are a number of different genetic polymorphisms that vary systematically among races, which, in interaction with environmental factors, result in different traits and behaviors (Walsh & Bolen 2012).

This is indeed what we consistently and robustly find. Allocco and his colleagues (2007) showed that data from just 50 randomly selected SNPs from 270 individuals was sufficient to predict their continent of origin with 95% accuracy. In another study, researchers were able to correctly assign 99 to 100% of individuals to their continent of origin using markers from only 100 genetic loci (Bamshad et al. 2003). A worldwide study using autosomal, mitochondrial, Y-chromosome, and *Alu*-insertion polymorphisms[3] concluded that their findings support "the current practice of grouping reference populations into broad ethnic categories" (Jorde et al. 2000:985). Between-population (African, Asian, European) variation was significantly different at <.0001 for all DNA analyses (Jorde et al. Table 4:982).

In yet another study, Tang and his colleagues (2005) used data from 326 genetic polymorphisms. Blind to their subjects' phenotypes, all individuals were assigned to racial/ethnic groups *a posteriori* using only DNA data. Using cluster analysis, only 5 out of a total of 3,636 (0.14%) subjects were not correctly classified according to their self-identified race. To correctly classify something that supposedly does not exist with such a small proportion of the available genetic markers with 99.86% accuracy belies those consider race to be a social fiction. Tang and his colleagues (2005:271) concluded that "The correspondence between genetic cluster and self-identified race/ethnicity [leads us to conclude that] major self-identified race/ethnicity and genetic cluster are effectively synonymous." If the no-race people require absolute 100% precision in a typology, they will undermine any definitional system.

The largest clustering study to date designed to discover if the human population clusters into identifiable subpopulations is that of Bamshad and his colleagues (2004). This study analyzed 63,724 SNPs in the regulatory and coding regions of 3,931 genes among 50,736 African and European Americans. They found 20,409 common SNPs (SNPs with a minor allele frequency of 10% in one or both races) only in African Americans and 2,802 common only in European Americans. Of these common SNPs, 23.1% were found only in African Americans and 2.9% only in European Americans. Less than half (48%) were common to both populations, and of these, 41% had significantly different allele frequencies between the two groups. Bamshad and colleagues further examined population structures using haplotypes (haplotypes—short for haploid genotype—are a group of alleles of different genes on a single chromosome of an individual that are closely linked and usually inherited as a unit). Only 51% of the identified haplotypes were shared by both groups. Thus, people who are commonly defined by race have patterns of polymorphism frequencies by which geneticists can distinguish them without having to resort to observing their physical features.

Conclusion

Numerous studies have affirmed a biological basis for what we know as race. With an adequate number of ancestry-informative markers, or AIMs, which are sets of polymorphisms for a locus known to have substantially different frequencies between people from different geographical regions, geneticists can even determine a person's ethnicity (e.g., a Spaniard from a German or even a Welshman from a Scot) with almost 100% accuracy (Paschou et al. 2010). Those who conduct this type of research may substitute more politically palatable terms for race, but all the substitutions that have been proposed retain the fundamental idea of biologically distinguishable subgroups of human beings. Fujimura and Rajagopalan (2010:11) argue that while these subgroups are demonstrably there, because the statistical "programs produce these clusters without any reference to self-reported race, ethnicity, or ancestry, they can be used to create categories of *genetic difference* that are *not* categories of race." This is a strange statement that seems to say that because no clusters were defined a priori with reference to race, the patterns defined a posteriori cannot be called race, even though they map almost perfectly to self-reported race.

Fujimura and Rajagopalan (2010:26) also complain that researchers often "slip into race statements." It is difficult to see how they can help doing so. With a wink and a nod, researchers know what they are looking at is the same phenomena the Enlightenment monogenists would recognize, albeit in a more nuanced and sophisticated way, regardless of under what rhetorical umbrella their research is framed. All such research (many hundreds of studies) has found an extraordinary degree of precision in genetic groupings that map to self-identified race. Researchers have never reached 100% success, but comparing the concordance between genes and self-reported sex and self-reported race, Neil Risch has pointed out that geneticists find "a higher discordance rate between self-reported sex and markers on the X chromosome [than between self-reported race and genetic markers]!" (Gitschier 2005:14). Some might think of this as a clincher argument in the debate, since no one doubts that male and female are natural kinds. Although there are fringe types who sometimes appear to claim that certain aspects of sex are social constructs, I don't believe that anyone has ever claimed that sex doesn't exist.

Endnotes

1.   The F_{st} statistic is a measure of population differentiation based the degree of heterozygosity in polymorphisms in two or more subpopulations relative to all subpopulations combined. The isolation of breeding populations increases homozygosity (the "fixation" of an allele in a breeding population). Gene flow between different breeding populations increases heterozygosity within population but decreases it between populations. Complete

subpopulation differentiation (no genetic overlap) would yield an F_{st} value of 1.0. The F_{st} statistic is calculated in the same way that the coefficient of determination (r^2) is calculated using the sum of squares approach: thus, $F_{st} = (Ht - Hs)/Ht$. Ht is the genetic variation within all populations, and Hs is the genetic variation within subpopulations.

2.   Lewontin a proud and unabashed Marxist and a shameless ideologue. He is a brilliant scientist, but he has no qualms whatsoever about making his science subservient to his ideology. Along with two fellow Marxists, he wrote, "We share a commitment to the prospect of the creation of a more socially just—a socialist—society" (Rose, Lewontin, & Kamin 1984: ix). To be fair, this was written before the fall of the iron curtain and the implosion of socialist states across Europe and elsewhere. Yet we continue to hear Marxists in academia yearn for a true Marxist society, arguing that the societies that imploded were corruptions of the real thing. Such folks seem to have an image of communism as a Platonic Form, with all those imploding states merely being distortions of the real thing. Those who yearn for secular salvation in the arms of Marx must ask themselves how many more social experiments featuring executions, gulags, and (ig)noble lies must we endure before they finally realize that the Form they idealize can never be instantiated in the phenomenal world. As Albert Einstein (and perhaps many others before him) once said, "The definition of insanity is doing the same thing over and over again and expecting different results."

3.   Mitochondrial DNA is non-nuclear DNA from mitochondria, which are organelles embedded in a cell and transmitted via the maternal line only, and Y chromosomal DNA is only transmitted via the paternal line. *Alu* insertions are short strands of DNA that replicate and insert themselves in different positions of the same chromosome or in different chromosomes. They typically have no effect on gene functioning.

Chapter 12

Is Peace Possible?

Know Thy Enemy

The science wars about the reality and legitimacy of so many things have raged for centuries and have provided endless grist for pundits of all stripes. Aristotelians and Platonists, with their different temperaments and visions, have fought each other about everything from the nature of the universe to the nature of human beings. Gender and race are only the most recent theaters of engagement. Unlike former times, we are most fortunate today to have the science and technology with which to quickly and directly engage contending issues. This remark once again betrays my bias in favor of the science side of the wars, but there are folks smarter than I who loathe and distrust the practice of science when it reveals inconvenient truths.

I have examined the character and spirit of science and social constructionism to provide a taste of what I believe they entail and why they tend to attract people of different visions or temperaments. The conceptual tools of science and the constructionists' objections to them were examined and oiled before taking them into the substantive fray. Unfortunately, it is evident that constructionists have made only the feeblest attempts to reconnoiter enemy positions before leading their followers into the conflict. Sun Tzu preached centuries ago that knowing your enemy is the first lesson of warfare, but few constructionists bother themselves to acquire even the rudiments of genetics, neuroscience, or evolutionary biology. One can never know the enemy as well as the enemy knows itself, but it behooves those who want to contend with them to know something of their strengths, as well as the weaknesses we believe them to have.

Some constructionists may have a grudging admiration and envy of science and have made efforts to study it, but they still find things in it to criticize. One criticism that looms large is that science cannot be value free. Because unconstrained social scientists approach issues of gender and race from an explicitly value-laden moral stance in their denials of meaningful race and gender differences, they assume that those who explore these differences do likewise, with science being just an elaborate smoke screen for racism and sexism. Since they have the moral high ground as anti-sexists and anti-racists, scientists who apply brain imaging machines, DNA centrifuges, and arcane statistical tools to examining these differences must be swimming in the moral swamp.

This reasoning confuses process with product. Subjectivity certainly enters the picture when scientists choose a topic to explore and the methods they will employ, but these value choices do not depreciate the products that issue from them. There may be sexists and racists engaged in this research, but there are researchers of all races and both sexes engaged in it, and it can hardly be said that they all are sexists and racists. The subjective processes that led the scientist to be interested, even passionately interested, in producing one result over another (no scientist wants the null hypothesis to win), has absolutely no bearing on the validity of the products of his or her endeavors. If the products pass the tests of peer appraisal, replication, and survive intact any *scientific* objections, the products can reasonably be said to be objective and value neutral.

Arguments from Fear

We have seen that many folks consider race to be a dangerous idea, and linking gender to biological sex is not exactly safe either. A remarkable number of arguments against studying gender and racial differences have been arguments from fear, a logical fallacy that philosophers call *argumentum ad metum.* They evidently work, because many who may have had interests in that directions have been convinced that exploring such things is a career buster as well as in loathsome taste. We have seen how the study of gender differences has been described as everything from neurosexism to rape. In terms of race, Richard Lewontin (1972:397) wrote many years ago that "Human racial classification is of no social value and is positively destructive of social and human relations." More up to date is Ash Amin's (2010) claim that "biological racism" has replaced "phenotypic racism" with the reinscription of race in terms of molecular genetics, and Cooper, Kaufman, and Ward's (2003:1169) assertion that "The discovery that races exist is not an advance of genomic science into uncharted territory, it is an extension of the atavistic believe that human populations are not just organized, but ordered." Hearing authoritative arguments like these would make anyone who wants to fit into academia as a good liberal to quake at the very mention of race. And finally, my primary guide in the thickets of constructionism, Ian Hacking, remarks that "the fear that all this DNA stuff will be put to racist purposes, including high-tech criminal profiling, is justified" (2006:87).

Why anyone would think something that assists the law to catch and convict murderers, rapists, and robbers (except of course, the murderers, rapist, or robbers) is a bad thing escapes me. One could have said the same thing about photography and fingerprints in the 19th and 20th centuries when they were novelties, because they also assisted the law to identify and convict miscreants. Further, saying that these things could be put to racist purposes implies that only blacks (and this is the race folks have in mind when they talk about racist purposes) commit nasty crimes and are dumb enough to leave their DNA lying around at the scene.[1] Let us not forget that DNA evidence has been used to clear people of crimes as well as convict them. Hacking goes on to say, however, that there is no hiding from the reality that DNA reveals that race is a reality, and believes that "it is quite possible that white liberals want to hide more than black Americans do." It certainly seems that blacks have far less fear of race than whites, since they are disproportionately represented among those who patronize commercial genetic ancestry companies both in the United States (Hartigan 2008) and the UK (Skinner 2006), and they may do what whites dare not: openly affirm their racial pride.

There are times when arguments from fear are legitimate when used to warn of real dangers. I want to scare the pants off everyone with worst-case scenarios about global warming and nuclear proliferation, because arguments of this type are meant to open a dialogue leading to calls for action. But what good are such arguments used against things that we cannot possibly do anything about, such as the existence of race and gender differences? All these arguments do is to contend that to uncover evidence of differences is to open a Pandora's Box from which will emerge old Jim Crow, the repeal of the Nineteenth Amendment granting women the vote, and many other such nefarious things. Of course, anything is possible, but to try to stifle important research that can help rather than harm those whom the censors seek to protect by evoking images of yesteryear is paranoia. Races and the genders will differ in their behavior and be noted by all who look, regardless of such arguments (perhaps even because of them, since they goad the curious into looking). David Skinner (2007:938) sums up my position exactly: "Sociology's repeated return to the critique of race as biology can only be justified if one believes that a revival of old style scientific racism is a genuine possibility. . . . I do not find this a plausible argument." Substitute gender for race and sexism for racism, and the same position applies.

The Moralistic Fallacy

Arguments from fear easily slip into the moralistic fallacy, the lesser-known sibling of the naturalistic fallacy. The naturalistic fallacy, a phrase coined by the Scottish philosopher David Hume, is the fallacy of leaping from *is* to *ought*. This is the tendency to believe that whatever is natural is good; in other words, no distinction is made between an empirical fact and a moral evaluation of it. Examples of the naturalistic fallacy would be to say that because violence is part of human nature (i.e., part of our evolved behavioral repertoire) we are saying that it is natural and therefore it is good or at least justifiable, or because men and women are different we are justified in treating them differently. Areas of the social sciences with an unabashed agenda, such as feminism, tend to be rife with the naturalistic fallacy (Buss & Malamuth 1996).[2]

The moralistic fallacy reverses the process to jump from *ought* to *is*; it deduces facts from moral judgments. Those who commit this fallacy argue that if something is morally offensive to them it is in fact false, or at least they encourage their value judgments to be treated as factually true. Thus, violence is a bad thing and therefore cannot be part of human nature, and men and women ought to be treated equally because there are no differences between the genders save in the plumbing. Lahn and Ebenstein (2009: 726) maintain that by promoting biological sameness, constructionists believe that discrimination against individuals or groups of individuals will be curtailed. The go on to say that this position, "although well-intentioned, is illogical and even dangerous, as it implies that if significant group diversity were established, discrimination might thereby be justified." As the 1951 UNESCO (1951:51) statement concluded on the matter, "We wish to emphasize that equality of opportunity and equality in law in no way depend, as ethical principles, upon the assertion that human beings are in fact equal in endowment."

Harvard biologist Bernard Davis supposedly coined the moralistic fallacy phrase in 1978 to counter censorial propositions being bandied about by latter-day Platonic guardians of allowable knowledge claiming that certain research topics should not be pursued (in this case, the genetics of intelligence) because it is "socially dangerous." Since these folks take it for granted that they have the monopoly on virtue, they feel that this monopoly grants them the right, indeed the duty, to conceal or destroy any knowledge contrary to their vision of social justice. Davis says that the notion of "forbidden knowledge" has a long history, but that "It is a difficult notion for scientists to accept, since all knowledge can be used in various ways, and it would seem better to restrain the bad uses rather than to deprive ourselves of the good ones" (2000:5). I think this is a position that almost all true scientists would hold.

Extreme political liberals are most likely to submit to the moralistic fallacy. Because most academics in the social sciences are liberals, there is a strong tendency to commit this fallacy when discussing many issues having to do with differences of gender and race (an example is Joan Roughgarden's dance around the facts of sexual selection theory discussed in Chapter 5). Moralizing is a necessary exercise in a civilized society for judging practices and behaviors as good or bad, but morality is not a legitimate criterion for making ontological statement about the facts of nature.

The insinuation of the morality into the realm of gender and race issues (as well as many other issues) helps us to understand why we so often see unconstrained visionaries viciously attacking constrained visionaries, but rarely do we see the opposite. It is the unconstrained vision's fusion of so many issues with morality that leads holders of that vision to consider those who disagree with them as immoral supporters of inequality rather than simply wrong-headed. As Thomas Sowell (1987:227) puts it, "Implications of bad faith, venality, or other moral and or intellectual deficiencies have been much more common in the unconstrained vision's criticism than vice versa." Unconstrained visionaries seem to believe that if apparently intelligent and well-educated individuals research race and sex differences, or oppose programs they believe could improve the lives of the less fortunate, they must be evil racists and sexists, or some other hissing epithet. On the other hand, constrained visionaries rarely accuse their opponents of deliberately opposing the common good, even if they see them as inadvertently doing just that. Constrained visionaries see their opponents as well-meaning but naive, wrong, or unrealistic; rarely do they resort to ad hominem attacks that impugn their moral characters.

155

Unconstrained visionaries tend to hold up Kant's deontological ethics whereby good intentions are in accordance with duty as the gold standard. Constrained theorists follow Jeremy Bentham's consequentialist ethics; they know that the intentions of unconstrained visionaries are good, perhaps even noble, but judge actions and policies as good or bad by consequences, not by intentions. As two trenchant old sayings have it "The road to hell is paved with good intentions," and "Sometimes the remedy is worse than the disease." [3]

Is Détente Possible?

The battle between biology and social constructionism does not have to be a zero-sum struggle ending in the unconditional surrender of either. We have seen that gender is not completely based on biology or on socialization, and that race is both biologically real and a social construct. The social constructionist side of the argument will have to make serious adjustments in its approach if its adherents want to engage biology. They can contest the evidentiary claims about the biological underpinnings of race and gender (or anything else), but it is not acceptable practice to assert dogmatically that there are none, or to use scare tactics to silence future research. Biologists will continue pouring out studies germane to race and gender, against which simple assertions of disbelief or ad hominem cries of sexism and racism will accomplish nothing other than laying bare the poverty of the constructionist argument. What constructionists must do, according to a sympathetic John Hardigan (2008:167), is to reformulate "assertions about the social construction of race by more sharply delineating what is cultural about race, rather than predicating these assertions on discrediting biological research." The exact same thing can be said, of course, for gender constructionists.

Hardigan's counsel is essentially that constructionists should retreat to their own borders and stay there. They can rest assured that the enemy will not pursue them into their territory, for there is nothing there that they want. Once back on home ground they can concentrate on examining socially constructed referents about race and gender, which must be, however, informed by the science pertaining to them or else the exercise will be futile. There are many interesting issues in this domain, and tending their own gardens was exactly what Berger and Luckmann (1966) advised us to do when they urged us to examine the social origins of our take-for-granted reality. Constructionists shoot themselves in the foot when they do otherwise. As Jenny Reardon

(2004:58), a feminist and friend of social constructionism, put it, "each statement about the biological meaninglessness of race acts to further enhance geneticists' control over its definition."

Gender, Race, and Interactive Kinds

If social scientists do come to the conclusion that there are indeed robust biological substrates of gender and race, what is left to cultivate in the constructionist garden about them? They might begin by noting Hacking's (1999) distinction between classifications in the natural sciences and classifications in the social sciences. Hacking demarcates the *indifferent* kinds of things classified in the former and the *interactive* kinds in the latter. The objects of the natural sciences are indifferent to how they are classified or treated (Pluto didn't give a damn when it lost its planetary status, and copper doesn't care if it is pounded into a prince's amulet or a peasant's chamber pot) and do not change in any way in response to the names humans call them. Of course, organisms of various kinds develop selection pressures in response to human interventions (microbial mutations due to antibiotics, for instance), and thus they react to human treatment, but they are obviously not aware of what they are doing (Hacking 1999:106).

Human beings, on the other hand, are not indifferent to their classifications or their treatment. Humans are interactive kinds who are fully aware of their classifications and of their treatment; that is, classifications interact with their objects of classification. This interaction is what Hacking called "looping effects," in which classification and the classified form feedback loops. The issue for social scientists is to explore the effects of classifications. In terms of gender and race, they have to look at referents about those classifications, what they symbolize and denote, not the classifications themselves, since gender or race are in-the-mirror givens (intersex and biracial individuals excepted). People can conform to, reject, or simply ignore the classifications of others, but Hacking would say that there are always some influences, however subtle, of classification schemes on persons being classified. Whatever these affects may be, they belong in the constructionist garden rather than mine, but I will briefly touch on self-esteem, since stereotypes (good or bad) are said to affect our sense of worthiness or unworthiness.

There are small male-female differences in global self-esteem, but when domain-specific self-esteem is examined, we see bigger gender difference, with females scoring higher than males in some instances, and males scoring higher than females in others. Numerous studies show that we locate our self-esteem primarily in evolutionarily relevant domains: males in such things as group status, athleticism, and sexual prowess, and females in terms of beauty and in the nature of their interpersonal relationships (Gentile et al. 2009). We also tend to base our sense of self-worth on our performance on specific things that are important to us and to ignore things about which we could not care less. The supposed negative stereotypes of females thus do not appear to have adversely affected female self-esteem.

Similarly, studies in the United States over the past 60 years or so have consistently found blacks to have significantly higher levels of global self-esteem than whites, who have significantly higher levels than Asians (see Table 1 in Chapter 11) (Zeigler-Hill 2007). This is the opposite of expectations based on the putative self-fulfilling nature of stereotypes. As discussed earlier, blacks have been the recipients of many negative stereotypes, while Asians have been the recipients of many positive stereotypes (model minority, smart, hard-working, law abiding). Self-esteem is domain and context-specific; blacks get it from things that are important to them, whites from what is important to them, and Asian from what is important to them. There are many biological, psychological, and social factors that influence self-esteem far removed from classificatory schemes or from stereotypes. Labels do have an impact, but they are hardly as powerful as they are alleged by some to be. Even if they were, the ethical battle should be against false imputation of inferiority/superiority and not against scientific studies of difference. The mission of critics of race and gender differences, should they chose to accept it, is to provide specific examples of how data on these differences are currently used against women and minorities, or may be in the future. Any such potential risks must then be weighed against claims of actual benefits made by medical and biological scientists, as well as some behavioral scientists exploring topics in which race is a salient factor, that accrue from recognizing these differences. They should then ask why it is not sensible to adopt Davis' (2000) advice to concentrate on restraining any bad uses of knowledge that may come to their attention rather than seeking to deny others access to that knowledge.

What are the benefits recognizing group or category differences to members of those groups or categories? It surely goes without saying that whatever differences exist, no good can ever come of denying them. Whatever the domain of interest may be—biology, medicine, social science, criminology—proceeding under a cloud of deliberate denial can only harm those whom we seek to help. Women had long complained that medicine was male oriented because most of the studies of disease risk and treatment had only male participants. The assumption was that, if this bias was indeed in evidence, except for reproductive functions males and females were alike, and the same risk, protective, and treatment factors thus applied equally to both sexes (Pinn 2009). Tuck Nygun and colleagues' (2011:240) review of the genetics of sex differences in the brain concluded that such evidence constitutes a boon to women: "As science continues to advance our understanding of sex differences, a new field is emerging focused on better addressing the needs of men and women: gender-based biology and medicine."

Similarly, medical science has assumed that race can be ignored when diagnosing and treating patients. Because of the influences of differences in genetic polymorphisms, the races differ in disease susceptibility and response to treatment just as the sexes do (Lahn & Ebenstein 2009; Risch 2006). Practicing physicians are well aware of this. In an op-ed piece in the *New York Times*, Sally Satel explained why she is a "racially profiling doctor": "In practicing medicine, I am not colorblind. I always take note of my patient's race. So do many of my colleagues. We do it because certain diseases and treatment responses cluster by ethnicity. Recognizing these patterns can help us diagnose disease more efficiently and prescribe medications more efficiently" (quoted in Hartigan 2008:169). She is not saying, of course, that race is the determining factor in shaping her treatment; she is only saying that it is an important piece of information in her medical evaluation of a patient.

A Rose Is a Rose by Any Other Name—Or Is It?

With the exception of *God* (and perhaps *free will*), never in the history of written language has so much ink been spilled over the origin, use, and meaning of a word, and the ontological status of what it designates, as the word *race*. The whole brouhaha over race seems to be more rhetorical than substantive, and is fogged over by the dark shadows of slavery and the Holocaust. We may have existed quite peacefully with the term and what it means had these abominations not occurred, but they did, and some scientists who like the concept but not the term, suggest that terms such as population or ethnicity (the AAA's favored term) replace race. We can call what we see anything we like (race, ethnicity, clines, ancestral markers, populations) or invent something entirely new. Nothing would be lost except a word if we did, but *race* is a term so embedded in colloquial usage that it almost seems perversely obscurantist to purge it from the lexicon. Scientists today have a different picture in their heads when they hear *race* than they did in previous generations, and the terms *population* and *ethnicity* carry other entrenched meanings on their backs.

In biology, the term *population* means a breeding population of individuals. In the distant past, populations were more or less synonymous with what would become races because breeding populations of *Homo sapiens* were isolated from one another in a sparsely populated planet cluttered with almost insuperable geographic barriers (mountain ranges, deserts, large bodies of water) separating them. Takahata (1995) notes that if human mating couples had been equally distributed around the world 100,000 years ago, there would have been only about 35 mating couples in the whole of the region we now call France. It was this relative isolation, of course, that was sufficient to diversify the genetic composition of the races. To quote Ernst Mayr (2002:91) again, "A human race consists of the descendants of a once-isolated geographical population primarily adapted for the environmental conditions of their original country." Based on mitochondria DNA (mtDNA) data, it has even been claimed that humans had already subdivided into races before moving out of Africa: "The existence of between-group differences [in mtDNA] far older than within-group differences implies that the late Pleistocene expansion of our species occurred separately in populations that had been isolated from each other for several tens of thousands of years" (Harpending et al. 1993:495).

Viewed from this cladistic (branching) perspective, race is a concept that helps us to explain many things in science. It helps us to understand human migration patterns, genetic reactions to adaptive pressures in different geographic and cultural environments, and the evolutionary history of our species. Surely this is a good and useful thing. Recall that it was the awareness of morphological differences among peoples that focused Enlightenment scientists on evolutionary thinking relative to how these differences came about, thus paving the way for Charles Darwin's magnificent theory.

It is reasonable to assume that very little gene flow occurred between populations until commerce and the vehicles for conducting it (domestication of large pack animals, the invention of the wheel and of sturdy sailing vessels) sent individuals into previous alien breeding populations beginning only about 4,000 years ago (Woodward 1992). The trickle soon became a flood as breeding populations become more accessible to one another, making gene flow an increasingly important process in human evolution. Given the increasing rate of genetic mingling in the modern world, race may indeed become a concept of purely historical interest in the distant future, but in the meantime it is still a reality.

As for ethnicity, in anthropology and sociology ethnicity means a person's national or regional cultural ancestry. A race is a much larger classificatory unit than a population or an ethnic group, subsuming many such classifications. Take Fred Schmidt, born in Saginaw, Michigan, to German immigrant parents. As commonly understood, his nationality is American, his ethnicity is German, his race is Caucasian, and his breeding population, more than likely, is in Saginaw, but depending on how much Fred gets around, is potentially anywhere in today's mobile world. If race is synonymous with population, there are millions of races; if it is synonymous with ethnicity, there are thousands. Clearly neither term is a sensible substitute for race. After all, Schmidt would still be Caucasian (a designation he shares with many ethnic groups) and ethnically German (shared only with people of German ancestry), but if his parents had migrated to any place on the planet from Australia to Zimbabwe, his relevant population and nationality would have changed.

Conclusion

The science wars over gender and race were fought for decades like the trench warfare of the First World War with very little ground being lost or gained by either side. Exponential increases in the neurological and genomic sciences over the last 25 years or so, however, have turned the various campaigns into something more akin to the blitzkrieg strategies of the Second World War than the bloody stalemates of the First. There is no getting away from the fact that literally thousands of studies from diverse disciplines using diverse methods have demonstrated numerous meaningful differences between the genders and the races.

Because "difference" implies terms of quantitative and qualitative judgment such as more, less, better, worse, superior, inferior, egalitarians wince at the term. Why the term should elicit such a response is something of a mystery, because diversity is what the world is about. Each gender and each race is, *on average*, superior to the other gender

161

or other races in some things, and inferior in others, just as each person is both superior and inferior to other individuals in any number of ways. What is so dreadfully wrong about honestly acknowledging the obvious? These different talents and abilities possessed in different average levels by the genders and races should be embraced and celebrated as the sources of evolutionary resilience rather than denounced and feared as sources of turmoil and hatred. I look forward to the day when only the silliest of Bruce Charlton's clever sillies believe that social justice can only be realized by denying average gender and racial differences.

But the final word belongs to race rather than gender, because it is race that generated the fiercest conflicts. Perhaps we should drop the term *race* for a less contentious term if by doing so we can get off the euphemistic treadmill. If we can move on to see what can be done in the social and biological sciences to advance the betterment of the *human race* by recognizing and celebrating the fact that we are different individually and racially, I will be happy to purge the word from my vocabulary. But if population and ethnicity are not suitable (at least in my opinion) candidates for replacement, what is? I agree with John Hartigan (2008), who favors the term *ancestry*, a term without a negative history and which maps to both geography and DNA.

While it does seem simplistic to simply change the word and hope that all the denotative baggage it carries will disappear with it, it is those who deny the reality of race who call for a terminological transition, not I. Whatever term may be eventually substituted for race, the human beings to whom it refers will still be biologically distinguishable subgroups. Geneticists consistently show with remarkable accuracy that humans can be grouped by small portions of their DNA into categories that fit the traditional racial categories, and it thus seems quixotic to deny those categories, regardless of the name we apply to them. These groupings are not essentialist discreet typologies with sharp boundaries, and there is no pattern of gene variants present in all individuals of one group and absent in all individuals of another group. Differences can be used for hateful purposes only if we allow them to be, and surely it is the purposes to which we put knowledge that should be monitored by social constructionists rather than battling the production of such knowledge. After all, it is not as though we can change the root of these differences, although we can change how we react to them. As Norman Levitt (1999:315) has concluded, "All of us, scientists and nonscientists alike, must ultimately create and sustain a society and a culture that is mature enough and brave enough to handle the gifts—and the uncomfortable truths—that science affords." To that I can only add the exclamation AMEN!

Endnotes

1. DNA evidence has been used to capture seven serial killers by the Los Angeles police department's cold case squad since 2004, four of whom were African American (two white, one Hispanic). These four men—Chester Turner, Michael Hughes, John Floyd Thomas, and Lonnie Franklin—were responsible for at least 57 rape/murders. Is their apprehension good (justice) or bad (racist)? At least half of their victims were black women; you decide (McGough 2011).

2. The signatories of the 1986 *Seville Statement on Violence* commit the naturalistic fallacy when they state that "biological findings have been used to justify violence and war" (in Fox 1988:36). The Statement provides no example of "biological findings" being used to "justify violence or war," because there are none. The signatories also commit the moralistic fallacy when they go on to write (in capital letters yet) that: "IT IS SCIENTIFI-CALLY INCORRECT to say that in the course of human evolution there has been a selection for aggressive behavior any more than for other kinds of behavior." Period, full stop, case closed! How they believed that the mechanisms of aggression and violence got into us is anyone's guess, and what could "any more than for other kinds of behavior" mean here? The qualifying phrase implies that the signatories knew that it has been selected for like "other kinds of behavior," but also that they want to wish that fact away.

3. According to Kant, ethics has both an empirical and rational components (Plato versus Aristotle one last time). The empirical component is practical morality whereby acts are judged good or bad according to their conse-quences (e.g., the honest shopkeeper who is honest solely because he does not want to drive his customers to the competition), but such acts are not moral per se, because they are done for practical rent-paying reasons. Only those acts performed solely from duty and good will (the rational component) are truly moral, and are adjudged so *regardless of the conse-quences*. Thus, policies such as Prohibition and the current drug war are moral because they were derived from good moral intentions (to prevent the use of mind-altering substances that often lead to violence), despite their negative consequences. For unconstrained visionaries, capitalism is immoral (at least amoral) despite the bounty it provides us because it is a system built on self-interest, which is not moral for Kant. The concern for constrained visionaries is that unconstrained visionaries are so concerned with intentions that they are oblivious to historical lessons, which all too often have led to horrible consequences.

References

Abu El-Haj, N. (2007). The genetic reinscription of race. *Annual Review of Anthropology*, 36:283–300.

Adkins, D., & G. Guo (2008). Societal development and the shifting influence of the genome on status attainment. *Research in Social Stratification and Mobility*, 26:235–255.

Alexander, G. (2003). An evolutionary perspective on sex-typed toy preferences: Pink, blue, and the brain. *Archives of Sexual Behavior*, 32:7–14.

Alexander, G., T. Wilcox, & R. Woods (2009). Sex differences in infants' visual interest in toys. *Archives of Sexual Behavior*, 38:427–433.

Allen, C. (2007). It's a boy! Gender expectations intrude on the study of sex determination. *DNA and Cell Biology*, 26:699–705.

Alcock, J. (2001). *The triumph of sociobiology.* New York: Oxford University Press.

Allocco, D., Q. Song, G. Gibbons, M. Ramoni, & I. Kohane (2007). Geography and genography: Prediction of continental origin using randomly selected single nucleotide polymorphisms. *BMC Genomics*, 8:68–76.

American Anthropological Association (1998). AAA statement on "race." *American Anthropologist*, 100:712–713.

Amin, A. (2010). The remainders of race. *Theory, Culture & Society*, 27:1–23.

Amunts, K., E. Armstrong, A. Malikovic, L. Homke, H. Mohlberg, A. Schleicher, & K. Zills (2007). Gender-specific left-right asymmetries in human visual cortex. *The Journal of Neuroscience*, 27:1356–1364.

Andersson, M., & L. Simmons (2006). Sexual selection and mate choice. *Trends in Ecology and Evolution*, 21:296–302.

Ansolabehere, S., & C. Stewart (2009). Amazing race: How post-racial was Obama's victory? *Boston Review*, January/February. Online at: http://www.bostonreview.net/BR34.1/ansolabehere_stewart.php.

Arnold, A. (2004). Sex chromosomes and brain gender. *Nature Reviews: Neuroscience*, 5:1–8.

Arnold, A., J. Xu, W. Grisham, X. Chen, & Y. Kim (2009). Minireview: Sex chromosomes and brain differentiation. *Endocrinology*, 145:1057–1062.

Ash, J., & G. Gallup (2007). Paleoclimatic variation and brain expansion during human evolution. *Human Nature*, 18:109–124.

Atkins, P. (2003). *Galileo's finger: The ten great ideas of science.* New York: Oxford University Press.

Babbitt, C., O. Fedrigo, A. Pfefferle, A. Boyle, J. Horvath, T. Furey, & G. Wray (2010). Both noncoding and protein-coding RNAs contribute to gene expression evolution in the primate brain. *Genome Biology and Evolution*, 2:67–79.

Badcock, C. (2000). *Evolutionary psychology: A critical introduction*. Cambridge, England: Polity Press.

Bailey, D., & D. Geary (2009). Hominid brain evolution: Testing climactic, ecological, and social competition models. *Human Nature*, 20:67–79.

Bamshad, M., S. Wooding, W. Watkins, C. Ostler, M. Batzer, & L. Jorde (2003). Human population genetic structure and inference of group membership. *American Journal of Human Genetics*, 72:578–589.

Bamshad, M., S. Wooding, B. Salisbury, & J. Claiborne Stephens (2004). Deconstructing the relationship between genetics and race. *Nature Reviews: Genetics*, 5:598–609.

Banton, M. (2010). The vertical and horizontal dimensions of the word race. *Ethnicities*, 10:127–140.

Barkow, J. (1992). Beneath new culture is an old psychology: Gossip and social stratification. In J. Barkow, L. Cosmides, & J. Tooby (eds.), *The Adapted mind: evolutionary psychology and the generation of culture.* (pp. 627–637). New York: Oxford University Press.

Bartels, A., & S. Zeki (2004). The neural correlates of maternal and romantic love. NeuroImage, 21: 1155–1166.

Bateman, P., & N. Bennett (2006). The biology of human sexuality: Evolution. ecology, and physicology, *Verbum et Ecclesia*, 27:245–264.

Beaver, K. (2009). Molecular genetics and crime. In A. Walsh & K. Beaver. *Biosocial criminology: New directions in theory and research*, pp. 59–72. New York: Routledge.

Bell, E., J. Schermer, & P. Vernon (2009). The origins of political attitudes and behaviors: An analysis using twins. *Canadian Journal of Political Science*, 42:855–879.

Benedict, R. (1942). *Race and racism*. London: Routledge and Kegan Paul.

Bennett, S., D. Farrington, & L. Huesman (2005). Explaining gender differences in crime And violence: The importance of social cognitive skills. *Aggression and Violent Behavior*, 10:263–288.

Bentley, C. (2005). Family, humanity, polity: Theorizing the basis and boundaries of political community in Frankenstein. *Criticism*, 47:325–351.

Berger, P., & T. Luckmann (1966). *The social construction of reality: A treatise in the sociology of knowledge.* Garden City, NY: Anchor Books.

Bishop, R. (2006). Determinism and indeterminism. In *Encyclopedia of Philosophy*, (2nd ed.), D. M. Borchert (ed.). Farmington Hills, MI: Macmillian Reference.

Blackburn, R. (2000). *The overthrow of colonial slavery: 1776–1848*. London: Verso.

Bobo, L., & J. Kluegel (1997). Status, ideology, and dimensions of whites' racial beliefs and attitudes: Progress and stagnation. In S. Tuch & J. Martin (eds.), *Racial attitudes in the 1990s: Continuity and change*, pp. 93–120. Westport, CT: Praeger.

Bohan, J. (1993). Regarding gender: Essentialism, constructionism, and feminist psychology. *Psychology of Women Quarterly*, 17:5–21.

Boghossian, P. (2006). *Fear of Knowledge: Against Relativism and Constructivism*. New York: Oxford University Press.

Bonham, V., E. Warhauer-Baker, & F. Collins (2005). Race and ethnicity in the genome era. *American Psychologist* 60:9–15.

Bouchard, T., N. Segal, A. Tellegen, M. McGue, M. Keyes, & R. Krueger (2003). Evidence for the construct validity and heritability of the Wilson-Patterson conservatism scale: A reared-apart twins study of social attitudes. *Personality and Individual Differences*, 34:959–969.

Boyle, M. (1990). *Schizophrenia: A scientific delusion*. London: Routledge.

Bradley, S., G. Oliver, A. Chernick, & K. Zucker (1998). Experiment of nurture: Ablatio penis at 2 months, sex reassignment at 7 months, and a psychosexual follow-up in young adulthood. *Pediatrics*, 102, 91–95.

Brebner, J. (2003). Gender and emotions. *Personality and Individual Differences* 34:387– 394.

Briken, P., N. Habermann, W. Berner, & A. Hill (2006). XYY chromosome abnormality insexual homicide perpetrators. *American Journal of Medical Genetics*, 141b:198– 200.

Bromage, T. (1987). The biological and chronological maturation of early hominids. *Journal of Human Evolution*, 16–257–272.

Buchanan, P. (2011). Obama's race-based spoils system. http://twonhall.com/columists/Patbuchana/2011/08/26.

Burnyeat, M. (1976). Protagoras and self-refutation in later Greek philosophy. The Philosophical Review, 85:44–69.

Buss, D. (2001). Human nature and culture. *Journal of Personality*, 68:955:978.

Buss, D., & N. Malamuth (1996). Introduction. In Sex, power, and conflict. Evolutionary and feminist perspectives. D. Buss & N. Malamuth (eds.), pp. 3–5. New York: Oxford University Press.

Byne, W. (2006). Developmental endocrine influences on gender identity. *The Mount Sinai Journal of Medicine*, 73:959–959.

Byrnes, J., D. Miller, & W. Schafer (1999). Gender differences in risk taking: A meta-analysis. *Psychological Bulletin*, 125:367–383.

Cahill, L., N. Uncapher, L. Kilpatrick, M. Alkire, & J. Turner (2004). Sex-related hemispheric lateralization of amygdala function in emotionally-influenced memory: An fMRI investigation. Learning and Memory, 11:261–266.

Campbell, A. (1999). Staying alive: Evolution, culture, and women's intrasexual aggression. *Behavioral and Brian Sciences*, 22:203–214.

Campbell, A. (2004). Female competition: Causes, constraints, content, and contexts. *The Journal of sex research*, 41:16–26.

Campbell, A. (2006). Feminism and evolutionary psychology. In J. Barkow (ed.), Missing the revolution: Darwinism for social scientists, pp. 63–99. Oxford: Oxford University Press.

Campbell, A. (2009). Gender and crime: An evolutionary perspective. In A. Walsh & K. Beaver (eds.), *Criminology and Biology: New directions in theory and research*, pp. 117–136. New York: Routledge.

Campbell, A., S. Muncer, & D. Bibel (2001). Women and crime: An evolutionary approach. *Aggression and Violent Behavior*, 6:481–497.

Campbell, A., L. Shirley, & J. Candy (2004). A longitudinal study of gender-related cognition and behaviour. *Developmental Science*, 7:1–9.

Canli, T. (2006). *Biology of personality and individual differences*. New York: Guilford.

Carter, B. (2000). *Realism and racism: Concepts of race in sociological research*. London: Routledge.

Cartwright, N. (1997). Why physics? In. R Penrose & M Longair (eds.), *The large, the small and the human mind*, pp. 161–168. Cambridge, England: Cambridge University Press.

Casswell, B. (2009). The presidency, the vote, and the formation of new coalitions. *Polity*, 41:388–407.

Cavalli-Sforza, L. (2000). *Genes, peoples, and languages*. New York: North Point Press.

Cavalli-Sforza, L., P. Menozzi, & A. Piazza (1994). *The history and geography of human genes*. Princeton, NJ: Princeton University Press.

Chapple, C., & K. Johnson (2007). Gender differences in impulsivity. *Youth Violence and Juvenile Justice*, 5:221–234.

Charlton, B. (2009). Clever sillies: Why high IQ people tend to be deficient in common sense. *Medical Hypotheses*, 73:867–870.

Cheng, Y., P. Lee, C-Y. Yang, C-Y. Lin, D. Hung, & J. Decety (*2008*). Gender differences in the *murhythm* of the human mirror-neuron system. *PloS One*, 3:e2113.

Chura, L., M. Lombardo, E. Ashwin, B. Auyeung, B. Chakrabarti, E. Bullmore, & S. Baron-Cohen (2010). Organizational effects of fetal testosterone on human corpus callosum size and asymmetry. *Psychoneuroendocrinology*, 35:122–132.

Cohen, Kettenis (2010). Psychosocial and psychosexual aspects of disorders of sex development. *Best Practices & Research Clinical Endocrinology and Sex Development*, 24:325–334.

Colapinto, J (2006). *As nature made him: The boy who was raised as a girl*. New York: Harper Collins.

Cole, M. (2009). *Critical race theory and education: A Marxist response*. London, Palgrave.

Cole, S. (ed.). (2001). *What's wrong with sociology?* New Brunswick, NJ: Transaction.

Collins, F. (2006). *The Language of God*. New York: Free Press.

Collins, H. (1981). Stages in the empirical programme of relativism. *Social Studies of* Science, 11:3–10.

Cooper, R., J. Kaufman, & R. Ward (2003). Race and genomics. *New England Journal of Medicine*, 348:1166–1170.

Corredoira, M. (2009). Quantum mechanics and free will: Counter-arguments. *NeuroQuantology*, 7:449–456.

Costa, P., A. Terracciano, & R. McCrae (2001). Gender differences in personality traits across cultures: Robust and surprising findings. *Journal of Personality and Social Psychology*, 81: 322–331.

Craig, I., E. Harper, & C. Loat (2004). The genetic basis for sex differences in human behaviour: Role of sex chromosomes. *Annals of Human Genetics*, 68:269–284.

Cronin, H. (2003) Getting human nature right. In J. Brockman (ed.), *The new humanist: science at the edge*, pp. 53–65. New York: Barnes & Noble.

Cummings, M. (2000). *Human heredity: principles and issues*. London: Brooks/Cole.

Curnoe, D., A. Thorne, & J. Coate (2006). Timing and tempo of primate speciation. *Journal of Evolutionary Biology*, 19:59–65.

Davies, A., & T. Shackleford (2008). Two human natures: How men and women evolved different psychologies. In C. Crawford & D. Krebs (eds.), *Foundations of evolutionary psychology*, pp. 261–281. Danvers, MA: CRC Press.

Davis, B. (2000). The scientist's world. *Microbiology and Molecular Biology Reviews*, 64:1–12.

Dawkins, R. (1998). Postmodernism disrobed. *Nature*, 394:141–143.

Dennett, D. (1995). *Darwin's Dangerous Idea: Evolution and the Meanings of Life.* New York: Simon & Schuster.

Derntl, B., A. Finkelmayer, S. Eickhoff, T. Kellerman, D. Falkenberg, F. Schnieder, & U. Habel (2010). Multidimensional assessment of empathetic abilities: Neural correlates and gender differences. *Psychoneuroendocrinology*, 35:67–82.

Depue, R., & P. Collins (1999). Neurobiology of the structure of personality: Dopamine, facilitation of incentive motivation, and extraversion. *Behavioral and Brain Sciences*, 22:491–569.

Desai, R., & H. Eckstein (1990). Insurgency: The of peasant rebellion. *World Politics*, 42:441–456.

deVries, G., & P. Sodersten (2009). Sex differences in the brain: The relation between structure and function. *Hormones and Behavior*, 55: 589–596.

de Waal, F. (2002). Evolutionary psychology: The wheat and the chaff. *Current Directions in Psychological Science*, 11:187–191. New York: Charles Scribner.

de Waal, F. (2008). Putting the altruism back into altruism: The evolution of empathy. *Annual Review of Psychology*, 59:279–300.

Diamond, M. (1999). Pediatric management of ambiguous and traumatized genitalia. *The Journal of Urology*, 162:1021–1028.

Diamond, M. (2009). Clinical implications of the organizational and activational effects of hormones. *Hormones and Behavior*, 55:621–632.

Dobzhansky, T. (1973). Nothing in biology makes sense except in light of evolution. *The American Biology Teacher*, 35:125–129.

Domes, G., M. Heinrichs, A. Michel, C. Berger, & S. Herpertz (2007). Ocytocin imporoves "mind-reading in humans. *Biological Psychiatry*, 61:731–733.

D'Sousa, D. (1995). *The end of racism: Principles for a multiracial society.* New York: The Free Press.

Dunbar, R. (2007). Male and female brain evolution is subject to contrasting selection pressures in primates. *BioMedCentral Biology*, 5:1–3.

Dunbar, R., & S. Shultz (2007). Evolution of the social brain. *Science*, 317:1344–1347.

Dunn, L. (1969). Statement on the nature of race and race differences. In *Four statements on the race question*, pp.36–43. Paris, France: UNESCO.

Duster, T. (2006). Comparative perspectives in competing explanations: Taking on the newly configured reductionist challenge to sociology. *American Sociological Review* 71:1–15.

Eagly, A., W. Wood, & M. Johannesen-Schmidt (2004). Social role theory of sex differences and similarities. In A. Eagely, A. Beall, & R. Sternberg (eds.), *Psychology of gender*, pp. 269–295. New York: Guilford Press.

Edwards, A. (2003). Human genetic diversity: Lewontin's fallacy. *BioEssays*, 25:798–801.

Ehrenreich, B., & J. McIntosh (1997). The new creationism: Biology under attack, *The Nation*, 9:12–16.

Eichner, H. (1982). The rise of modern science and the genesis of romanticism. *PMLA*, 97:8–30.

Einstein, A. (1923). *Sidelights on Relativity (Geometry and Experience).* New York: P. Dutton.

El Hamel, C. (2002). "Race", slavery and Islam in Maghribi Mediterranean theowght: The question of the Haratin in Morocco. *The Journal of North African Studies*, 7:29–52.

Eliaeson, S. (2002). *Max Weber's methodologies: Interpretation and critique.* Malden, MA: Blackwell Press.

Esch, T., & G. Stefano (2005). Love promotes health. *Neuroendocrinology Letters,* 3:264– 267.

Evans, P., S. Gilbert, N. Mekel-Bobrov, E. Vallender, J. Anderson, L. Vaez-Azizi, S. Tishkoff, R. Hudson, & B. Lahn (2005). *Microcephalin,* a Gene Regulating Brain Size, Continues to Evolve Adaptively in Humans. *Science,* 309:1717–1720.

Farrar, W. (1964). Sir B. C. Brodie and his calculus of chemical operations. *Chymia,* 9:169–179.

Fausto-Sterling, A. (2002). Gender identification and assignment in Intersex Children. *Dialogues in Pediatric Urology* 25:4–5.

Figueredo, A., G. Vasquez, B. Brumbach, & S. Schneider (2007). The K factor, covitality, and personality: A psychometric test of life history theory. *Human Nature,* 18:47–73.

Figueredo, A., B. Vasquez, B. Brumbach, S. Schneider, J. Sefcek, I. Tal, & W. Jacobs (2006). Consilience and life history theory: From genes to brain to reproductive strategy. *Developmental Review,* 26: 243–275.

Fine, C. (2010). *Delusions of gender: How our minds, society, and neurosexism create difference.* New York: WW Norton.

Fine, C., & J. Kennett (2004). Mental impairment, moral understanding and criminal responsibility: Psychopathy and the purpose of punishment. *International Journal of Law and Psychiatry,* 27:425–443.

Fisher, H., A. Aron, & L. Brown (2005). Romantic love: An fMRI study of a neural mechanism for mate choice. *The Journal of Comparative Neurology,* 493: 58–62.

Fortune, W. (1939). Apapesh warfare. *American Anthropologist,* 41:22–41.

Foster, M. (2009). Looking for race in all the wrong places: Analyzing the lack of productivity in the ongoing debate about race and genetics. *Human Genetics,* 126:355–362.

Freedman, D. (1997). Is nonduality possible in the social and behavioral sciences? Small essay on holism and related issues. In N. Segal, G. Weisfeld, & C. Weisfeld (eds.), *Uniting Psychology and Biology,* pp. 47–80. Washington, DC: American Psychological Association.

Fromm, H. (2006). Science wars and beyond. *Philosophy and Literature,* 30:589–589.

Fuentes, A., J. Marks, T. Ingold, R. Sussman, P. Kirch, E. Brumfiel, R. Rapp, F. Ginsburg, L. Nader, & C. Kottak (2010). On nature and the human. *American Anthropologist,* 112:512–521.

Fujimura, J., & R. Rajagopalan (2010). Different difference: The use of "genetic ancestry" versus race in biomedical human genetics. *Social Studies of Science,* 41:5–30.

Gannett, L. (2010). Questions asked and unasked: How by worrying less about the "really real" philosophers of science might better contribute to debates about genetics and race. *Synthese,* 177:363–385.

Garrett, B. (2009). *Brain and behavior: Introduction to biological psychology.* Los Angeles: Sage.

Galernter, J., H. Kranzler, & J. Cubells (1997). Sewrotonin transporter protein (SLC6A4) allele and haplotype frequencies and linkage disequilibria in African- and European-American and Japanese populations in alcohol-dependent subjects. *Human Genetics,* 101:243–246.

Geary, D. (2000). Evolution and proximate expression of human paternal investment. *Psychological Bulletin*, 126:55–77.

Geary, D. (2005). *The origin of mind: Evolution of brain, cognition, and general
intelligence*. Washington, DC: American Psychological Association.

Gemmingen, M., B. Sullivan, & A. Pomerantz (2003). Investigating the relationships between boredom proneness, paranoia, and self-consciousness.
Personality and Individual Differences, 34:907–919.

Gentile, B., S. Grabe, B. Dolan-Pascoe, & A. Maitino (2009). Gender differences
in domain-specific self-esteem: A meta-analysis. *Review of General Psychology*, 13:34–45.

Gergen, K. (1988). Feminist critique of science and the challenge of social epistemology. In M. McCanney Gergen (ed.), *Feminist thought and the structure
of knowledge* (pp. 27–48). New York: New York University Press.

Gewertz, D. (1981). A historical reconsideration of female dominance among the
Chambri of Papua New Guinea. *American Ethnologist*, 8:94–106.

Gewertz, D., & F. Errington (1991). *Twisted Histories, Altered Contexts: Represent-
ing The Chambuli in a World System.* Cambridge: Cambridge University Press.

Gintis, H. (2003). The hitchhiker's guide to altruism: Gene-culture coevolution and the internalization of norms. *Journal of Theoretical Biology*,
220:407–418.

Gitschier, J. (2005). The whole side of it—An interview with Neil Risch. *PLoS
Genetics*, I:3–5.

Goldberg, S. (1986). Utopian yearning versus scientific curiosity. *Society,* 23:29–39.

Gooren, L. (2006). The biology of human psychosexual development. *Hormones
and Behavior*, 50:589–601.

Gould, S., & N. Eldredge (1977). Punctuated equilibria: The tempo and mode of
evolution reconsidered. *Paleobiology,* 3:115–151.

Graves, J. (2001). *The emperor's new clothes: Biological theories of race at the
millennium.* New Brunswick, NJ: Rutgers University Press.

Gray, J., & P. Thompson (2004). Neurobiology of intelligence: Science and ethics.
Nature Reviews: Neuroscience, 5:471–482.

Greene, J. (1954). Some early speculations on the origin of human races. *American
Anthropologist*, 56:31–41.

Griskevicius, V., A. Delton, T. Robertson, & J. Tybur. (2011). Environmental
contingency in life history strategies: the influence of mortality and socio-
economic status on reproductive timing. *Journal of Personality and Social
Psychology*, 100:241–254.

Groves, C. (2004). The what, why and how of primate taxonomy. *International
Journal of Primatology*, 25:1105–1128.

Grosvenor, P. (2002). Evolutionary psychology and the intellectual left. *Perspectives
in Biology and Medicine*, 45:433–448.

Grusec, J., & P. Hastings (eds.) (2007). Introduction. *Handbook of socialization:
Theory and research,* pp. 1–9. New York: Guilford Press.

Gross, P., & N. Levitt (1994). *Higher superstition: The academic left and its quarrels
with science.* Baltimore, MD: Johns Hopkins University Press.

Grosvenor, P. (2002). Evolutionary psychology and the intellectual left. *Perspectives
in Biology and Medicine*, 45:433–448.

Gunnar, M., & K. Quevedo (2007). The neurobiology of stress and development.
Annual Review of Psychology, 58:145–173.

Gur, R. C, F. Gunning-Dixon, W. Bilker, & R. E. Gur (2002). Sex differences in temporo-limbic and frontal brain volumes of healthy adults. *Cerebral Cortex*, 12:998–1003.

Hacking, I. (1999) *The Social Construction of What?* Cambridge, MA: Harvard University Press.

Hacking, I. (2006). Genetics, biosocial groups & the future of identity. *Daedalus*, 135: 81–95.

Harry, D., & J. Marks (1999). Human population genetics versus the HGDP. *Politics and the Life Sciences*, 18:303–305.

Hartigan, J. (2008). Is race still socially constructed? The recent controversy over race and medical genetics. *Science as Culture*, 17:163–193.

Hartman, A. (2004). The rise and fall of whiteness studies. *Race & Class*, 46:22–38.

Hannaford, I. (1996). *Race. The history of an idea in the West*. Baltimore: Johns Hopkins Press.

Harding, S. (1980). The norms of social inquiry and masculine experience. *Proceedings of the Biennial Meeting of the Philosophy of Science Association*, 2:305–324.

Harding, S. (1986). *The science question in feminism*. Ithaca, NY, and London: Cornell University Press.

Harding, S. (1991). *Whose science? Whose knowledge? Thinking from women's lives*. Ithaca, NY: Cornell University Press.

Hare, L., P. Bernard, F. Sanchez, P. Baird, E. Vilain, T. Kennedy, & V. Harley (2009). Androgen receptor repeat length polymorphism associated with male-to-female transsexualism. *Biological Psychiatry*, 65:93–96.

Harpending, H., & P. Draper (1988). Antisocial behavior and the other side of cultural evolution. In T. Moffitt, & S. Mednick (eds.), *Biological contributions to crime causation*, pp. 293–307. Dordrecht: Martinus Nyhoff.

Hapending, H., S. Sherry, A. Rogers, & M. Stoneking (1993). The genetic structure of ancient human populations. *Current Anthropology*, 34:483:496.

Harris, J., & C. Davidson (2009). Obama: The new contours of power. *Race and Class*, 50:1–19.

Harry, D., & J. Marks (1999). Human population genetics versus the HGDP. *Politics and the Life Sciences*, 18:303–305.

Hassett, J., E. Siebert, & K. Wallen (2008). Sex differences in rhesus monkey toy preferences parallel those of children. *Hormones and Behavior*, 54:359–364.

Hawks, J., E. Wang, G. Cochran, H. Harpending, & R. Moyzis (2007). Recent acceleration of human adaptive evolution. *Proceedings of the National Academy of Science*, 104:20753–20758.

Hepper, P. (2005). Unravelling our beginnings. *The Psychologists*, 18:474–477.

Hermans, E., P. Putman, & J. van Honk (2006). Testosterone reduces empathetic mimicking in healthy young women. *Psychoneuroendocrinology*, 31, 859–866.

Hermens, D., L. Williams, I. Lazzaro, S. Whitmont, D. Melkonian, & E. Gordon (2004). Sex differences in adult ADHD: A double dissociation in brain activity and autonomic arousal. *Biological Psychology*, 66:221–233.

Herschbach, D. (1996). Imaginary gardens with real toads. *Annals of the New York Academy of Sciences*, 775:11–30.

Hines, M. (2004). *Brain gender*. Oxford: Oxford University Press.

Hines, M. (2006). Prenatal testosterone and gender-related behavior. *European Journal of Endocrinology*, 155:115–121.

Hines, M. (2011). Gender Development and the Human Brain. *Annual Review of Neuroscience*, 34: 69–88.

Hines, M., & G. Alexander (2008). Monkeys, girls, boys and toys: a confirmation letter regarding "Sex differences in toy preferences: striking parallels between monkeys and humans." *Hormones and Behavior*, 54:359–364.

Hines M., S. Golombok, J. Rust, K. J. Johnston, J. Golding, & the Avon Longitudinal Study of Parents and Children Study Team (2002). Testosterone during pregnancy and gender role behavior of preschool children: a longitudinal, population study. *Child Development*, 73:1678–1687.

Hublin, J., & H. Coqueugniot (2006). Absolute or proportional brain size: That is the question. *Journal of Human Evolution*, 50:109–113.

Hudson, N. (1996). From "nation" to "race:" The origin of racial classification in eighteenth-century thought. *Eighteenth-Century Studies*, 29:247–264.

Hughes, M. (1997). Symbolic racism, old fashioned racism, and whites' opposition to Affirmative action. In S. Tuch & J. Martin (eds), *Racial attitudes in the 1990s: Continuity and change,* pp. 45–75. Westport, CT: Praeger.

Jagger, A (1986). Love and knowledge: Emotion in feminist epistemology. In A. Jagger & S. Bordo (eds.), *Gender/body/knowledge: Feminist reconstructions of knowing and being,* pp. 145–171. New Brunswick: Rutgers University Press.

Jausovec, N., & K. Jausovec (2008). Spatial rotation and recognizing emotions: Gender related differences in brain activity. *Intelligence*, 36:383–393.

Jeffrey, L. (1962). Wordsworth and science. *The South Central Bulletin*, 27:16–22.

Jonason, P. B. Koenig, & J. Tost (2010). Living a fast life: The Dark Triad and life history theory. *Human Nature*, 21:428–442.

Jorde, L., W. Watkins, M. Barnshad, M. Dixon, C. Ricker, M. Seielstad, & M. Batzer (2000). The Distribution of human genetic diversity: A comparison of mitochondrial, autosomal, and Y-chromosome data. *American Journal of Human Genetics*, 66:979–988.

Jost, J., & D. Amodio (2011). Political ideology as motivated social cognition: Behavioral and neuroscience evidence. *Motivation and Emotion.* Doi 10.1007/s11031–011–9269–7.

Jurgensen, M., O. Hiort, P. Holterhus, & U. Thyen (2007). Gender role behavior in children with XY karyotype and disorders of sex development. *Hormones and Behavior*, 51:443–453.

Kaiser, J. (2003). African-American population biobank proposed. *Science*, 300:1445.

Kalderon, M. (2009). Epistemic relativism. *Philosophical Review*, 118:225–240.

Kanazawa, S. (2008). Temperature and evolutionary novelty as forces behind the evolution of general intelligence. *Intelligence*, 36:99–108.

Kanai, R., T. Feilden, C. Firth, & G. Rees (2011). Political orientations are correlated with brain structure in young adults. *Current Biology*, 21:677–680.

Kaszyka, K., G. Strkalj, & J. Strzatko (2009). Current views of European anthropologists on race: The influence of education and ideological background. *American Anthropologist*, 111:43–56.

Kaszycka, K., & J. Strzalko (2003). "Race"—still an issue for physical anthrology? Results from Polish studies seen in the light of U.S. Findings. *American Anthropologist*, 105:116–124.

Kay, B. (2006). Blaming whitey. *http://proudtobecanadian.ca/index/writergroup/5436.*

Kelly, A. (2009). "I just tell the bloody truth, as I see it": James Kelman's *A Disaffection*, the Enlightenment, Romanticism and Melancholy Knowledge. *Études écossaises*, 12. http://etudesecossaises.revues.org/index193.html.

Kennair, L. (2002). Evolutionary psychology: An emerging integrative perspective within the science and practice of psychology. *The Human Nature Review*, 2:17–61.

Kenrick, D., & J. Simpson (1997). Why social psychology and evolutionary psychology need one another. In J. Simpson & D. Kenrick (eds.), *Evolutionary Social Psychology*, pp. 1–20. Mahwah, NJ: Lawrence Erlbaum.

Kimura, D. (1992). Sex differences in the brain. *Scientific American*, 267:119–125.

Klein, D. (1995). The etiology of female crime: A review of the literature. In B. Price & N. Sokoloff (eds.), *The criminal justice system and women: Offenders, victims, and workers* (pp. 30–53). New York: McGraw-Hill.

Klein, J., & N. Takahata (2002). *Where do we come from? The molecular evidence of human descent.* Berlin: Springer-Verlag.

Klinman, R., & N. Johnson (2005). What every undergraduate should know about evolution (and why). *Bioscience*, 55:926–928.

Knickmeyer, R., S. Baron-Cohen, P. Raggatt, K. Taylor, & G. Hackett (2006). Fetal Testosterone and empathy. *Hormones and Behavior*, 49:282–292.

Konigsberg, L., B. Algee-Hewitt, & D. Steadman (2009). Estimation and evidence in forensic anthropology: Sex and race. *American Journal of Physical Anthropology*, 139:77–90.

Kochanska, G., & A. Knaack (2003). Effortful control as a personality characteristic of young children: Antecedents, correlates, and consequences. *Journal of Personality*, 71:1087–1112.

Kohn, M. (1999). A race apart. *Index on Censorship*, 28:79–83.

Kolchin, P. (2002). Whiteness studies: The new history of race in America. *The Journal of American History*, 89:154–173.

Kraemer, B., T. Noll, A. Delsignore, G. Milo, U. Schnyder, & U. Hepp (2009). Finger length ratio (2D:4D) in adults with gender identity disorder. *Archives of Sexual Behavior*, 38:359–363.

Kuhn, T. (1970). *The structure of scientific revolutions.* Chicago: University of Chicago Press.

Kuhn, T. (1977). *The Essential Tension: Selected Studies in Scientific Tradition and Change.* Chicago: University of Chicago Press.

Lahn, B., & L. Ebenstein (2009). Let's celebrate human genetic diversity. *Nature*, 46:726–728.

Laks, A. (1990). Legislation and demiurgy: On the relationship between Plato's Republic and Laws. *Classical Antiquity*, 9:209–229.

Landau, I. (1998). Feminist criticism of metaphors in Bacon's philosophy of science. *Philosophy*, 73:47–61.

Latour, B. (2004). Why has critique run out of steam: From matters of fact to matters of concern. *Critical Inquiry*, 30:225–248.

Leslie, C. (1990). Scientific racism: Reflections on peer review, science and ideology. *Social Science and Medicine*, 31:891–912.

Lester, B., E. Tronick, E. Nestler, T. Abel, B. Kosofsky, C. Kuzawa, C. Marsit, I. Maze, M. Meaney, L. Monteggia, J. Reul, D. Skuse, J. Sweatt, & M. Wood (2011). Behavioral epigenetics. *Annals of the New York Academy of Sciences*, 1226:14–33.

Levine, D. (2006). Neural modeling of the dual motive theory of economics. *The Journal of Socio-Economics*, 35:613–625.

Levitt, N. (1999). *Prometheus bedeviled: Science and the contradictions of contemporary culture.* New Brunswick, NJ: Rutgers University Press.

Lewis, B. (1990). *Race and slavery in the Middle East.* New York: Oxford University Press.

Lewontin, R. (1972). The apportionment of human diversity. *Evolutionary Biology*, 6:391–398.

Lieberman, P. (1984). *The biology and evolution of language.* Cambridge, MA: Harvard University Press.

Lieberman, L., & R. Kirk (2002). The 1999 status of the race concept in physical anthropology: Two studies converge. *American Journal of Physical Anthropology*, Suppl.34:102.

Lindenfors P. (2005). Neocortex evolution in primates: the 'social brain' is for females. *Biology Letters*, 1:407–410.

Lindenfors, P., C. Nunn, & R. Barton (2006). Primate brain architecture and selection in relation to sex. *BioMedCentral Biology*, 5:1–9.

Lippa, R. (2003). *Gender, nature, and nurture.* Mahwah, NJ: Lawrence Erlbaum.

Lober, J. (1994). *The paradoxes of gender.* New Haven, CT: Yale University Press.

Lopreato, J., & T. Crippen (1999). *Crisis in sociology: The need for Darwin.* New Brunswick, NJ: Transaction.

Loury, G. (1995). *One by one from the inside out. Essays and reviews on race and responsibility in America.* New York: Free Press.

Luders, E., K. Narr, E. Zaidel, P. Thompson, L. Jancke, & A. Toga (2006). Parasagital asymmetries of the corpus callosum. *Cerebral Cortex*, 16:346–354.

Lynn, M. (1989). Race differences in sexual behavior: A critique of Rushton and Bogaert's evolutionary hypothesis. *Journal of Research in Personality*, 23:1–6.

MacEachern, S. (2006). Africanist archaeology and ancient IQ: Racial science and cultural evolution in the twenty-first century. *World Archaeology*, 38:72–92.

Manica, A., & R. Johnstone (2004). The evolution of paternal care with overlapping broods. *The American Naturalist*, 164:517–530.

Marks, J. (1996). Science and race. *American Behavioral Scientist.* 40:123–133.

Marx, K. (1967). *Capital* (volume 1). New York: International Publishers.

Marx, K. (1978). Economic and philosophical manuscripts of 1844, pp. 66–123, *The Marx-Engels reader*, R. Tucker (ed.). New York: W.W. Norton.

Mayr, E. (2002). The biology of race and the concept of equality. *Daedalus*, 131:89–94.

McCrae, T., & A. Terracciano (2005). Universal features of personality traits from the observer's perspective: Data from 50 cultures. *Journal of Personality and Social Psychology*, 88:547–561.

McGough, M. (2011). Not forgotten. *Miller-McCune*, November/December, 54–65.

McIntyre, M., & C. Edwards (2009). The early development of gender differences. *Annual Review of Anthropology*, 38:83–97.

Mead, M. (1935). *Sex and temperament in three primitive societies.* New York: Morrow.

Mead, M. (1949). *Male and female: A study of the sexes in a changing world.* New York: Morrow.

Mealey, L. (1990). Differential use of reproductive strategies by human groups? *Psychological Science*, 1:385–387.

Mekel-Bobrov, N., S. Gilbert, P. Evans, E. Vallender, J. Anderson, R. Hudson, S.Tishkoff, & B. Lahn (2005). Ongoing adaptive evolution of *ASPM*, a brain size determinant in *Homo sapiens*. *Science*, 309:1729–1722.

Mellon, R. (2007). A field guide to social construction. *Philosophy Compass*, 2:93–108.

Mendonca, B., M, Inacio, E. Costa, A. Maria Frade, J. Ivo, D. Russell, & J. Wilson (2003). Male pseudohermaphroditism due to 5[alpha]-reductase 2 deficiency: Outcome of a Brazilian cohort. *The Endocrinologist*, 13:201–204.

Merchant, C. (1980). *The death of nature: Women, ecology and the scientific revolution*. New York: HarperCollins.

Merten, J. (2005). Culture, gender and the recognition of the basic emotions. *Psychologia*, 48:306–316.

Meyer-Bahlburg, H. (2005). Gender identity outcome in female-raised 46, XY persons with penile agenesis, cloacal exstrophy of the bladder, or penile ablation. *Archives of Sexual Behavior*, 34:423–438.

Meyer-Bahlburg, H., C. Dolezal, S. Baker, A. Ehrhardt, & M. New (2006). Gender development in women with congenital adrenal hyperplasia as a function of disorder severity. *Archives of Sexual Behavior*, 35:667–684.

Mikkola, M. (2008). Feminist perspectives on sex and gender. *Stanford Encyclopedia of Philosophy*. Online at http://plato.stanford.edu/entrie/feminism-gender/.

Miller, E. (1994). Tracing the genetic history of modern man. *Mankind Quarterly*, 35:71–108.

Mitchell, K. (2007). The genetics of brain wiring: From molecule to mind. *PLoS Biology*, 4:699–692.

Mitton, J. (1977). Genetic differentiation of race of man as judged by single locus and multilocus analysis. *The American Naturalist*, 111:202–212.

Money, J. (1986). *Venuses Penuses: Sexology, Sexosophy, and Exigency Theory*. Buffalo, NY: Prometheus.

Montague, B. (1841). *The works of Francis Bacon, vol. III*. Philadelphia: Casey and Hart.

Moore, M. (2004). *Stupid white men—and other sorry excuses for the state of the nation!* New York: HarperCollins.

Morning, A. (2007). "Everyone knows it's a social construct": Contemporary science and the nature of race. *Sociological Focus*, 40:436–454.

Morning, A. (2008). Reconstructing race in science and society: Biology textbooks, 1952–2003. *American Journal of Sociology*, 114 Suppl. S106–S137.

Nassiry, M., A. Javanmard, & Reza Tohidi (2009). Application of statistical procedures for analysis of genetic diversity in domestic animal populations. *American Journal of Animal and Veterinary Sciences*, 4:136–141.

Nicholson, L. (1994). Interpreting gender. *Signs*, 20:79–105.

Neubauer, A., & A. Fink (2009). Intelligence and neural efficiency. *Neuroscience and Biobehavioral Reviews*, 33:1004–1023.

Norden, J. (2007). Understanding the brain. Chantilly, VA: The Teaching Company.

Norenzayan, A., & S. Heine (2005). Psychological universals: What are they and how can we know? *Psychological Bulletin*, 131:763–784.

Ngun, T., N. Ghahramani, F. Sanchez, S. Bocklandt, & E. Vilain (2011). The genetics of sex differences in brain and behavior. *Frontiers in Neuroendocrinology*, 32:227–246.

O'Brien, G. (2006). Behavioural phenotypes: Causes and clinical implications. *Advances in Psychiatric Treatment*, 12:338–348.

Oderberg, D. (2011). Essence and properties. *Erkenn*, 75:85–111.

Okasha, S. (2002). *Philosophy of science: A very short introduction*. Oxford: Oxford University Press.

Olson, J., P. Vernon, & J. Harris (2001). The heritability of attitudes: A study of twins. *Journal of Personality and Social Psychology*, 80:845–860.

Owen, T. (2006). Genetic-social science and the study of human biotechnology. *Current Sociology*, 54:897–917.

Parsons, L, & D. Osherson (2001). New evidence for distinct right and left brain systems For deductive versus probabilistic reasoning. *Cerebral Cortex*, 11:954–965.

Paschou, P., J. Lewis, A. Javed, & P. Drineas (2010). Ancestry informative markers for fine-scale individual assignment to worldwide populations. Journal of Medical Genetics, 47:835–847.

Pasterski, V., M. Geffner, C. Brain, P. Hindmarsh, C. Brook, & M. Hines (2005). Prenatal hormones and postnatal socialization by parents as determinants of male-typical toy play in girls with congenital adrenal hyperplasia. *Child Development*, 76:264–267.

Patten, M. (2010). Null expectations in subspecies diagnosis. *Ornithological Monograms*, 67:35–41.

Perry, B. (2002). Childhood experience and the expression of genetic potential: What childhood neglect tells us about nature and nurture. *Brain and Mind*, 3:79–100.

Pinker, S. (2002). *The blank slate: The modern denial of human nature*. New York: Viking.

Pinn, V. (2003). Sex and gender in medical studies. *Journal of the American Medical Association*, 289:397–400.

Plavcan, J., & C. van Schaik (1997). Intrasexual competition and body weight dimorphism in anthropoid primates. *American Journal of Physical Anthropology*, 103:37–68.

Plomin, R., K. Ashbury, & J. Dunn (2001). Why are children in the same family so different? Nonshared environment a decade later. *Canadian Journal of Psychiatry*, 46:225–233.

Plomin, R., J. Defries, I.Craig, & P. McGuffin (2003). Behavioral genomics. In R. Plomin, J. Defries, I. Craig, & P. McGuffin (eds.), *Behavioral genetics in the post-genomic era*, pp. 531–540. Washington, DC: American Psychological Association.

Polsek, D. (2009). Who has won the science wars? *Drustvena Istrazivanja (Journal of General Social Science)*, 18:1023–1047.

Porteous, A. (1934). Platonist or Aristotelian? *The Classical Review*, 48:97–105

Pratt, T., & F. Cullen (2000). The empirical status of Gottfredson and Hirschi's general theory of crime: A meta-analysis. *Criminology*, 38:931–964

Preston, J., & C. Chadderton (2012). Rediscovering 'Race Traitor': A critical race theory informed by public pedagogy. *Race Ethnicity and Education*, 15:85–100.

Quartz, S., & T. Sejnowski (1997). The neural basis of cognitive development: A constructivist manifesto. *Behavioral and Brain Sciences*, 20:537–596.

Quinlan, R., & M. Quinlan (2007). Evolutionary ecology of human pair bonds: Cross cultural tests of alternative hypotheses. *Cross-Cultural Research*, 41:149–169.

Qvarnstrom, A., J. Brommer, & L. Gustafsson (2006). Testing the genetics underlying the co-evolution of mate choice and ornament in the wild. *Nature*, 44: 84–86.

Radhakrishna, B. (2009). Galileo Galilei (1564–1642): The 400th anniversary of the invention of the telescope causing a great revolution in astronomy. *Journal of the Geological Society of India*, 74:429–432.

Reardon, J. (2004). Decoding race and human difference in the genomic age. *Differences: A Journal of Feminist Cultural Studies*, 15:38–65.

Richerson, P., & R. Boyd (2010). *Evolution since Darwin: The first 150 years.* In M. Bell, D. Futuyma, W. Eanes, & J. Levinton (eds.), Sunderland, MA: Sinauer, pp. 561–588.

Ridley, M. (1999). *Genome: the autobiography of a species in 23 chapters.* New York: HarperCollins.

Ridley, M. (2003). *Nature via nurture: Genes, experience and what makes us human.* New York: Harper Collins.

Risch, N. (2006). Dissecting racial and ethnic differences. *New England Journal of Medicine*, 354:408–411.

Robertson, W. (1997). Abolitionists of whiteness. *Instauration*, 22:12.

Roll-Hansen, N. (1984). E. S. Russell and J. H. Woodger: The failure of two twentieth-century opponents of mechanistic biology. *Journal of the History of Biology*, 17:399–428.

Romaine, C., & C. Reynolds (2005). A model of the development of frontal lobe functioning: Findings from a meta-analysis. *Applied Neuropsychology*, 12:199– 201.

Rorty, R. (1991). *Objectivity, Relativism and Truth: Philosophical Papers I.* Cambridge: Cambridge University Press, 1991.

Rosario, V. (2009). Quantum sex: Intersex and the molecular deconstruction of sex. *GLO: A Journal of Lesbian and Gay Studies*, 15:267–284.

Roscoe, P. (2003). Margaret Mead, Reo Fortune, and Mountain Apapesh warfare. *American Anthropologist*, 105:581–591.

Rose, S. (1999). Precis of *Lifelines:* Biology, freedom, determinism. *Behavioral and Brain Sciences*, 22:871–921.

Rose, S. (2001). Moving on from old dichotomies: Beyond nature-nurture towards a lifeline perspective. *British Journal of Psychiatry*, 178:3–7.

Rose, S., R. Lewontin, & J. Kamin (1984). *Not in Our Genes: Biology, Ideology, and Human Nature.* New York: Viking Penguin.

Rossi, A. (1984). Gender and Parenthood. *American Sociological Review*, 49:1–19.

Roughgarden, J. (2009). *The genial gene: Deconstructing Darwinian selfishness.* Berkeley: University of California Press.

Rowe, D. (1994). *The limits of family influence: Genes, experience, and behavior.* New York: Guilford Press.

Rushton, J. (1991). Race differences: A reply to Mealey. *Psychological Science*, 2:126.

Rushton, J. (1994). The equalitarian dogma revisited. *Intelligence*, 19:263–280.

Rushton, J. (2000). *Race, evolution, and behavior: A life history perspective* (2nd abridged ed.). Port Huron, MI: Charles Darwin Research Institute.

Rushton, J, (2011). Life history theory and race differences: An appreciation of Richard Lynn's contribution to science. *Personality and Individual Differences*, doi:10.1016/j.paid.2011.

Sampson, R., & W. J. Wilson (2000). Toward a theory of race, crime, and urban inequality. In S. Cooper (ed.), *Criminology*, pp. 149–160. Madison, Wisconsin: Courewise.

Sagan, C. (1980). *Cosmos.* New York: Random House.

Saint-Amand, P. (1997). Contingency and enlightenment. *SubStance,* 26:96–109.

Sarich, V., & F. Miele (2004). *Race: The reality of human differences. Boulder,* CO: Westview.

Sarukkai, S. (2005). Revisiting the "unreasonable effectiveness" of mathematics. *Current Science,* 88:415–423.

Sayer, A. (1997). Essentialism, social constructionism, and beyond. *The Sociological Review,* 45:452–487.

Sayers, S. (2005). Why work? Marx and human nature. *Science & Society,* 69:606–616.

Schmitt, D., A. Realo, M. Voracek, & J. Allik (2008). Why can't a man be more like a woman? Sex differences in big five personality traits across 55 cultures. *Journal of Personality and Social Psychology,* 94:168–182.

Schon, R., & M. Silven (2007). Natural parenting—back to basics in infant care. *Evolutionary Psychology,* 5:102–183.

Schulte-Ruther, M., H. Markowitsch, G. Fink, & M. Piefke (2007). Mirror neuron and theory of mind mechanisms involved in face-to-face interactions: A functional magnetic resonance imaging approach to empathy. *Journal of Cognitive Neuroscience,* 19:1354–1472.

Sesardic, N. (2010). Race: A social destruction of a biological concept. *Biology & Philosophy,* 25:143–162.

Shah, B. (2009). A challenge to "malestream" determinism. http://www.docstoc.com/docs/12115436/WOMENS-WRITING.

Silverman, I., J. Choi, & M. Peters (2007). The hunter-gatherer theory of sex differences in spatial abilities: Data from 40 countries. *Archives of Sexual Behavior,* 36:261–268.

Singer, P. (2000). *A Darwinian left: Politics, evolution, and cooperation.* New Haven, Yale University Press.

Sisk, C., & J. Zehr (2005). Pubertal hormones organize the adolescent brain and behavior. *Frontiers in Neuroendocrinology,* 26:163–174.

Skinner, D. (2007). Groundhog day: The strange case of sociology, race, and "science." *Sociology,* 4:931–943.

Smith, K., D. Oxley, M. Hibbing, J. Alford, & J. Hibbing (2011). Linking genetics and political attitudes: Reconceptualizing political ideology. *Political Psychology,* 32:369–397.

Sokal, A. (1996). Transgressing the boundaries: Toward a transformative hermeneutics of quantum gravity. *Social Text,* 46/47:217–252.

Sokal, A. (1996). A physicist experiments with cultural studies. *Lingua Franca,* May/June: 62–64.

Sokal, A., & J. Bricmont (1998). *Fashionable nonsense: Postmodern intellectuals' abuse of Science.* New York: Picador.

Sowell, T. (1987). *A conflict of visions: Ideological origins of political struggles.* New York: William Morrow.

Smith, N. (1983). Aristotle's theory of natural slavery. *Phoenix,* 37:109–122.

Snow, C. (1964). *The two cultures and a second look.* New York: Cambridge University Press.

Spiro, M. (1975). *Children of the Kibbutz.* Cambridge, MA: Harvard University Press.

Spiro, M. (1980). *Gender and culture: Kibbutz women revisited.* New York: Schocken.

Spiro, M. (1999). Anthropology and human nature. *Ethos*, 27:7–14.

Steele, S. (1991). *The content of our character*. New York: HarperCollins.

Takahata, N. (1995). A genetic perspective on the origin and history of humans. *Annual Review of Ecology and Systematics*, 26:343–372.

Tang H., T. Quertermous, B. Rodriguez, S. Kardia , X. Zhu, A. Brown, J. Pankow, M. Province, S. Hunt, E. Boerwinkle, N. Schork, & N. Risch (2005) Genetic structure, self-identified race/ethnicity, and confounding in case-control association studies. *American Journal of Human Genetics*, 76:268–75.

Templeton, A. (1998). Human races: A genetic and evolutionary perspective. *American Anthropologist*, 100:632–650.

Terranova, A., A. Morris, & P. Boxer (2008). Fear reactivity and effortful control in overt and relational bullying: A six-month longitudinal study. *Aggressive Behavior*, 34:104–115.

Thernstrom, S., & A. Thernstrom (1997). *America in black and white: One nation indivisible*. New York: Simon and Schuster.

Thorne, B. (1993.) Gender play: Girls and boys in school. New Brunswick, NJ: Rutgers University Press.

Tinbergen, N. (1963) On aims and methods in ethology. *Zeitschrift für Tierpsychologie*, 20:419–433.

Tooby, J., & L. Cosmides (1992). The psychological foundation of culture. In J. Barkow, L. Cosmides, & J. Tooby (eds.), The adapted mind: Evolutionary psychology and the generation of culture, pp. 19–136. New York: Oxford University Press.

Trautner, H. (1992). The development of sex-typing in children. *German Journal of Psychology*, 16:183–199.

Trigg, R. (1999). *Ideas of human nature: An historical introduction*. Oxford: Blackwell.

Trudge, C. (1999). Who's afraid of genetic determinism? *Biologist*, 46:96.

Twardosz, S., & J. Lutzker (2010). Child maltreatment and the developing brain: A review of neuroscience perspectives. *Aggression and Violent Behavior*, 15:59–68.

Udry, R. (1994). The nature of gender. *Demography*, 31:561–573.

UNESCO (1969). *Four statements on the race question*, pp. 36–43. Paris, France: UNESCO.

U.S. Equal Employment Opportunity Commission (2009). Court Services and Offender Supervision Agency. http://www1.eeoc.gov//federal/reports/fsp2009/csosa>cfm?renderforprint=1.

U.S. Office of Personnel Management (2006). *Demographic profile of the federal workforce*. http://www.opm.gov/feddata/demograp/demograp.asp.

Unger, R. (1996). Using the master's tools: Epistemology and empiricism. In S. Wilkinson (ed.), *Feminist social psychologies: International perspectives*, pp. 165–181. Milton Keynes, England: Open University Press.

van As, A., G. Fieggen, & P. Tobias (2007). Sever abuse of infants–an evolutionary price for human development? *South African Journal of Children's Health*, 1:54–57.

Vandermassen, G. (2004). Sexual selection: A tale of male bias and feminist denial. *The European Journal of Women's Studies*, 11:9–26.

van den Berghe, P. (1990). Why most sociologists don't (and won't) think evolutionarily. *Sociological Forum*, 5:173–185.

Vickers, B. (2008). Francis Bacon, feminist historiography, and the dominion of nature. *Journal of the History of Ideas*, 69:117–141.

Wakefield, J. (2000). Aristotle as sociobiologist: The "function of a human being" argument, black box essentialism, and the concept of mental disorder. *Philosophy, Psychiatry and Psychology*, 7:17–44.

Walsh, A. (2009). *Biology and criminology: The biosocial synthesis*. New York: Routledge.

Walsh, A. (2011). *Feminist criminology through a biosocial lens*. Durham, NC: Carolina Academic Press.

Walsh, A., & J. Bolen (2012). *The neurobiology of criminal behavior: Gene-brain-culture interaction*. Farnham, England: Ashgate.

Walsh, A., & C. Hemmens (2011). *Law, justice, and society: A sociolegal introduction*. New York: Oxford University Press.

Wang, Q., G. Strkali, & L. Sun (2003). On the concept of race in Chinese biological anthropology: Alive and well. *Current Anthropology*, 44:403.

Washington, B. (1972). *The Booker T. Washington papers*, Vol. 1. L. Harlan (ed.). Chicago: University of Chicago Press.

Watters, E. (2006). DNA is not destiny. *Discover: Science, Technology and the Future*. November.

Wells, B. (1980). *Personality and heredity*. London: Longman.

West, M. (2005). *The Poverty of Multiculturalism*, London: Civitas.

Williams, L., M. Barton, A. Kemp, B. Liddell, A. Peduto, E. Gordon, & R. Bryant (2005). Distinct amygdala-autonomic arousal profiles in response to fear signals in healthy males and females. *NeuroImage*, 28:618–626.

Williams, W. (2002). The devil made me do it. http://www.mugu.com/cgi.bin/ Upstream.

Wilson, E. O. (1998). *Consilience: The unity of knowledge*. New York: Alfred A. Knopf.

Wilson, W. J. (1987). *The truly disadvantaged*. Chicago: University of Chicago Press.

Woo, L., J. Thomas, & J. Brock (2010). Cloacal exstrophy: A comprehensive review of an uncommon problem. *Journal of Pediatric Urology*. April.

Wood. W., & A. Eagly (2002). A cross-cultural analysis of the behavior of women and men: Implications for the origins of sex difference. *Psychological Bulletin*, 128: 699–727.

Woodley, M. (2010). Is Homo sapiens polytypic? Human taxonomic diversity and its implications. *Medical Hypothesis*, 74:195–201.

Woodward, V. (1992). *Human heredity and society*. St. Paul, MN: West Publishing.

Wright, J., & P. Boisvert (2009). What biosocial criminology offers criminology. *Criminal Justice and Behavior*, 36:1228–1239.

Wu M., D. Manoli, E. Fraser, J. Coats, J. Tollkuhn, S-I. Honda, N. Harada, & N. Shah (2009). Estrogen masculinizes neural pathways and sex-specific behaviors. *Cell*, 139:61–72.

Xu, J., & C. Disteche (2006). Sex differences in brain expression of X- and Y-linked genes. *Brain Research*, 1126:59–55.

Yang, J., L. Baskin, & M. DiSandro (2010). Gender identity in disorders of sex development: Review article. *Urology*, 75:153–159.

Yee, A., H. Fairchild, F. Weizmann, & G. Wyatt (1993). Addressing psychology's problems with race. *American Psychologist*, 48:1132–1140.

Zack, N. (2002). *Philosophy of science and race.* New York: Routledge.

Zeigler-Hill, V. (2007). Contingent self-esteem and race: Implications for the black self-esteem advantage. *Journal of Black Psychology,* 35:51–74.

Zucker, K. (2002). Intersexuality and gender identity differentiation. *Journal of Pediatric Adolescent Gynecology,* 15:3–13.

Zuckerman, M. (2007). *Sensation seeking and risky behavior.* Washington, DC: American Psychological Association.

Zuriff, G. (2002). Inventing racism. *The Public Interest,* winter:114–128.

Index